Ultimate Parkinson's Tips

Ultimate Parkinson's Tips

to
Walk Faster, Stand Up, Unfreeze,
Turn in Bed, and More
(2nd Edition)

It's About Tricking the Neural Pathway

Graham O'Connor

Published by Graham O'Connor
2nd Edition: March 2020

First Printing: 2020

ISBN 978-0-359-71880-1

Graham O'Connor
8 Smolkin Street
Arnprior, Ontario, Canada K7S 3R9

oconnorgd@gmx..com

Dedication

To all the Parkinson's suffers.

I hope this book helps make your life a little easier.

Table of Contents

Ultimate Parkinson's Tips

Introduction

This book is about techniques I use to improve mobility when Parkinson's Disease (PD) causes slowness, initiation of movement, and freezing to become a problem. I'm in mid-stage PD. I shuffle, walking toe-to-heel with short steps. My left arm is difficult to swing and typing is difficult to do because Parkinson's is in both hands but especially hard with my left hand where my original tremor began. My balance has weakened though I have fallen down yet. When meds are working perfectly with no side-effects which is very infrequent I move well but most the time movement is compromised now. Except that I discovered techniques to improve muscle movement which allow me to reduce medication and reduce medication side-effects as a result. I have actively used these techniques with success. They will not make you normal again but they will make you move a little better and a little smoother. They will help you stand or get out of bed. They will help you quickly unfreeze if for example, you freeze suddenly while crossing the street. They can also make it so that you do not freeze while crossing the street in the first place, because you will be in the process of using techniques to move better that I believe help widen the neural pathway while the techniques are in use. I'll also touch on diet and exercise and other topics that are helpful to PwP (People with Parkinson's).

I'll get right to the point and keep things short and explain near the end in more detail how I came up with these ways to trick our Parkinson's bodies into better motion. I like getting to the point and providing the solutions in few words so that you don't need to read lengthy paragraphs. There are lots of photographs to help you understand the techniques. If at your current stage of PD you are now experiencing the particular symptom I cover, my technique should help, you will see. We people with Parkinson's have to work harder to move and we move less well. The techniques will help make movements easier, though not perfect. You will feel more confident to walk in public.

When you are slow or having difficulty moving, the techniques I will show increase your speed probably 30 to 40%, in general, while you actively use the technique. When I walk outside or climb stairs, sometimes I feel that it is higher than a 40% increase. While it is difficult to quantify the percentage, there is definitely noticeable improvement. You'll feel more confident in public even though you'll need to multitask by using techniques to move about. If you are having Dystonia or Dyskinesia moments the techniques are less effective as those side-effects take on a life of their own. You probably avoid public places when those side-effects happen. I practice taking low doses of meds to reduce med induced side-effects and the techniques help me stay on lower dosages.

The premise of the techniques that I use is that strong sensory sensation helps temporarily improve messages traveling through the neural pathways of our body. So if you are experiencing a slow, shuffling, toe-to-heel gait then the messages sent from your brain to your legs has weakened due to loss of dopamine. My theory is that the ever increasingly decreasing dopamine produced by our dying dopamine producing brain cells gradually results in the narrowing of the neural pathways due to less traffic in general from reduced dopamine. If, for example, a laser pointer creates a target on the floor for our eyes then the visual neural pathway is stimulated and this results in temporarily widening all pathways, allowing better messages to travel through our body, in this case helping improve our gait.

There are many sensory techniques that can be used, some better or more convenient for certain situations. And there are many applications. Things like walking, stairs, standing up, dressing, showering, typing, improving digestion, improving bowel movements, better balance, exercise and improving exercise, turning in bed, and much more are covered in the chapters to follow.

The senses that I found help improve Parkinson's symptoms are these three senses: Visual, Auditory, and Feel. Visual is what the eyes can see. Auditory is in the realm of the ears. Feel involves touch and muscle movement.

I'll begin in Chapter 2 by describing a test to see if these techniques will help you. Then I'll continue in Chapter 3 by listing techniques that you should use, which can be applied to various situations which I'll cover afterwards in Chapter 4 onwards.

I don't always think to use the techniques. I'll be shuffling and suddenly realize and start using a technique. Or I'll be showering and struggling and then pick a technique or two. Sometimes my meds are helping well enough on a good day and I don't need to use the techniques, though suddenly the off-time occurs and I use them.

Now, straight into the focus of this help book...

Chapter 2

Test If Sensory Techniques Will Work for You

This book is for People with Parkinson's who are at mid-stages of Parkinson's, that is, stage 2, 3, and 4. In other words, if you are at the stage where slow movement, initiation of movement, and shuffling gait has become a problem this book can help you get some mobility and muscle control back.

Basically if one technique works for you, then likely you will benefit from the rest of the techniques. A good test is with a laser pointer, Figure 2-1. If you do not have a laser pointer the use a flashlight in a room that is darkened enough so that you can easily see the light on the floor. Note if using the flashlight, the room the room doesn't have to be pitch black, you just need to be able to distinguish that there is a spot of light on the floor.

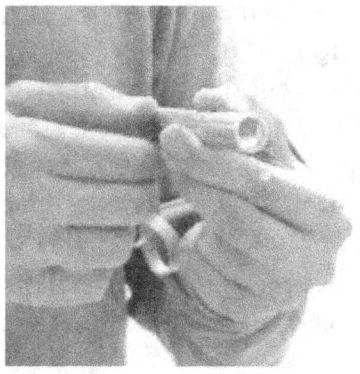

Figure 2- 1

To best see the result, try this test at a point during the day when you are walking slowly and shuffling. Maybe this is when you have just woken up and meds aren't working fully yet. Maybe it is when your meds are wearing off. Or possibly it is in the evening. Find the ideal time. For me the ideal time is the majority of the day as I practice low dosing my meds to drastically reduce medication side-effects.

For the test, with the laser pointer (or flashlight) turned off and in hand, stand at one end of the room, then begin walking. Be aware of how you are stepping. After several steps, turn the laser pointer (or flashlight) on and point it at the floor, several feet in front of you while continuing to walk forward. Keep walking and while you don't need to stare directly at the spot on the floor where the light is you should be aware of it, keeping it within your line of sight. You should notice that your stride changes, becomes smoother and more heel-to-toe. If this happens then you will benefit from the many tips in this book.

The tips go beyond using a laser pointer device and I cover much more than improving your gait

Chapter 3

The Neural Pathway and the Senses that will Help

The neural pathway

I envision the neural pathway to be like plumbing pipes that flows throughout our body. This is not a scientific analysis but instead is a metaphor to help visualize.

In a normal person's body, the brain produces adequate dopamine for sending messages through the pathways. The pathway pipe is steady. As we people with Parkinson's lose dopamine producing brain cells, we send weaker messages through the neural pathways. Over time, with less or smaller messages going through the 'pipe', the pipe narrows and messages to walk, to move, to use muscles are hampered by less dopamine but also by a narrowed neural pathway, Figure 3-1. By using sensory neural pathway tricks, extra messages go to the pathway, temporarily widening it and allowing clearer messages to get through. This results in a smoother gait, better movement, etc.

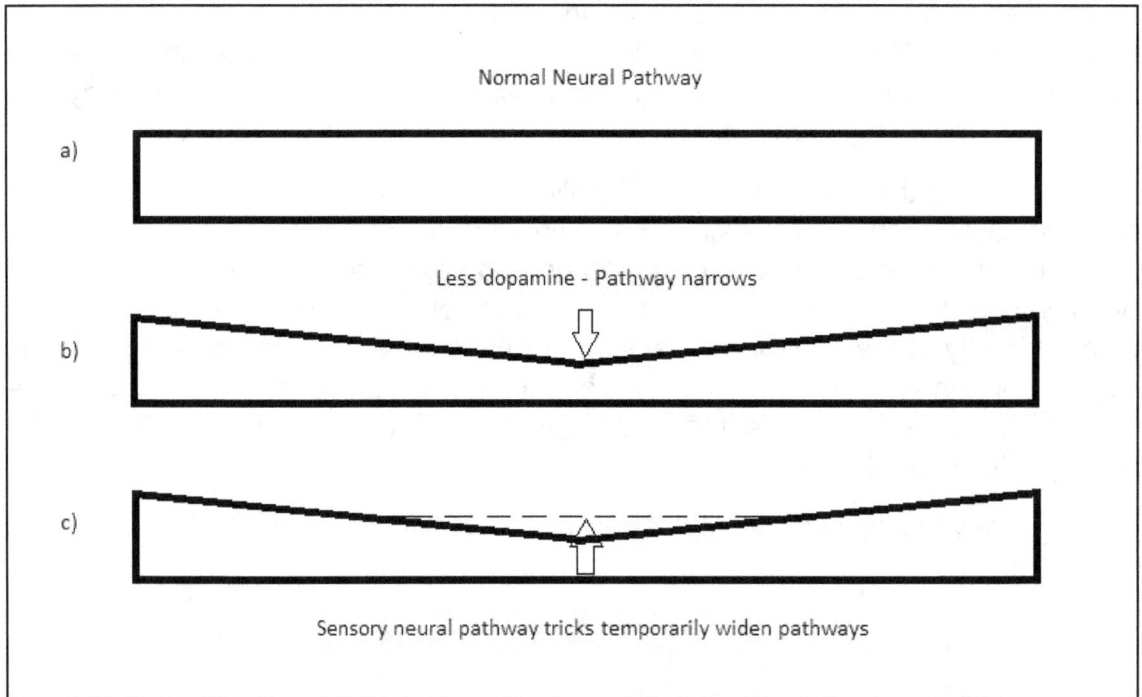

Figure 3- 1

The senses that help

I identified four senses that can be employed to temporarily widen the neural pathways and provide as much as 30 to 40% improvement in Parkinson's symptoms for the duration of time the neural pathway sensory technique is in use.

So if for example you walk 60% as good as you used to, employing a sensory trick may improve your gait by 60% * 40% = 24%. So you will temporarily improve to 60% + 24% = 84%.

If at 40% then you might improve to 40% * 40% = 16%. So you will temporarily improve to 40% + 16% = 56%.

We may not go back to 100% (temporarily) but the improvement is noticeable. The percentage of improvement is based on my observation.

The four senses involved:

1. Muscle
2. Touch
3. Sight
4. Hearing

Muscle

To stimulate the muscle sense moving/using certain muscles can activate the sensory trick.

Touch

Touch seems to be more of a combination of touch and muscle movement.

Sight

Sight can be engaged in a way that helps temporarily widen the neural pathways which in turn always all messages to get through better.

Hearing

The right sound will have a positive impact on Parkinson's symptoms.

Chapter 4

The Basic Techniques

What I have found is that when adding a simple additional movement in your body or stimulating certain senses, it is like the neuro pathway is temporarily opened wider and allows the message to the nerves to get through to the body part that is frozen, having trouble initiating movement, or slowed.

Basic techniques that will be used

Here are the techniques which you can refer back to as you go through this book.

Laser pointer technique

A laser pointer device shines creating a target for the eyes to instinctively focus on, acting as a target. This stimulates the visual neural pathway. A red laser light is ideal for indoors, nighttime, and any time the amount of daylight allows the red light to be easily visible. A more powerful green light laser pointer devices is best suited for bright daylight hours.

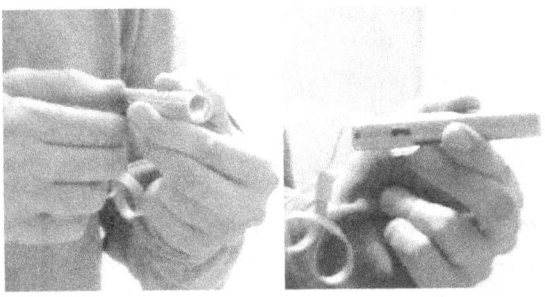

Figure 4- 1

You don't need to stare directly at the spot of light created on the ground, floor, wall, etc. and in the case of the green you should not stare directly at it.

You hold the laser in one hand while you make use of its benefits making sure to point the laser where it is close to centre of your vision so that you can be aware of it visually, Figure 4-1.

A headgear mounted laser light is great. It allows your hands to be free, Figure 4-2. You now have the option to use other neural pathway sensory techniques with your hands simultaneously. The laser light should be adjustable so that you can point it at the ground or a wall depending on the need. For example, if doing a punching bag workout you would point the laser at the wall and the bag.

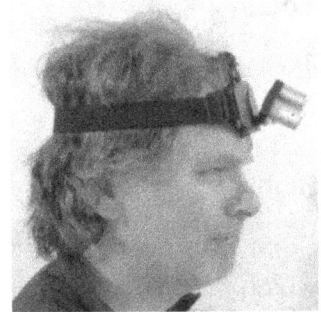

Figure 4- 2

Music with a strong beat technique

The Auditory Pathway can be used by sensory stimulation of the ears. Music with a strong beat, Figure 4-3, can be played to help with movement. I listen to Electronica and House music to get this effect. You can play it out loud, off your laptop, for example, or use your MP3 and earphones. It just depends on the situation and convenience.

Just the sound of a constant drum beat will affect the auditory sense in a way that helps Parkinson's mobility. The constant beat should be quick enough to stimulate the sense of hearing. Your entire brain is used when you listen to music.

Figure 4- 3

Bouncing a ball

While walking bounce a ball, Figure 4-4, and catch with one hand while the other rests in your jacket pocket or bounce back and forth between hands or some variation like that. You should see that your walking steps become smoother, longer, possibly quicker, and more properly heel to toe. This happens because the ball acts as a visual target, we instinctively chase it, we're actively using muscles to chase it trigger muscle sense, and the palm feels the ball.

Figure 4- 4

If using a racquetball ball you might like occasionally to bounce it a little harder and slightly more ahead of you, which may have the effect of automatically speeding up your walking pace as you chase the ball. You may wish to bounce it hard enough so that it bounces ahead of you and more than once. You will see that you automatically adjust your pace, speeding up or slowing down. Once you get a rhythm going and your timing you will see you walk more smoothly.

The ball: Use a racquetball ball, Figure 4-5, which is my preferance, or a tennis ball. Anything smaller than a racquetball ball might have its bounce direction affected by pavement cracks and stones and you will spend time chasing the ball which suddenly went unpredictably off course. A racquetball has more bounce and is better if the pavement is wet. I often carry the raquetball in my jacket pocket, the thigh pocket of my army pants, or in a fanny pack.

Figure 4- 5

Bouncing a ball while walking will give you confidence to walk further. You will also keep flexibility and dexterity in your arms as well as providing them with exercise. Multitasking keeps your brain sharp too. Why it works? I am not a scientist but in my opinion bouncing the ball activates messages through the neural pathway which temporarily widens the neural pathway allowing clearer messages to get through to the legs so that you walk smoother, faster, and heel-to-toe. Senses involved are muscle movement as the hand throws and catches the ball, the sense of touch, and a visual target as your eyes follow the ball's trajectory. There is also an instinct to move at necessary speed to be able to get to the ball before it begins bouncing too low.

Tossing a ball

You will toss the ball up a few inches with one hand or be-tween hands while you walk, Figure 4-6. You should see that your walking steps become smoother, longer, possibly quicker, and more properly heel to toe. The one hand toss can be done while carrying something in the other hand or while the other hand holds onto the stair railing.

Figure 4- 6

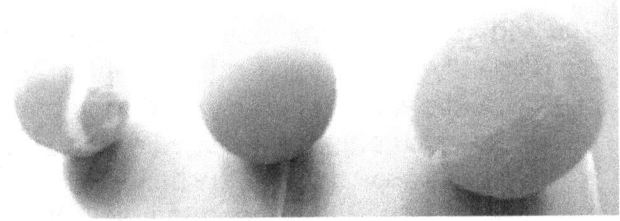

Figure 4- 7

Any ball can be used, from a tennis ball to anything smaller, Figure 4-7. If indoors, I prefer to use a smaller ball as the ball is mainly used for tossing.

I believe the secret to why tossing the ball works is not only the activeness of your hand but also that your palm is facing upwards to hold, toss, and catch the ball. It's all about increasing the neural messaging pathway.

Shaking a ball

Cup your hand and sort of shake your hand side to side rapidly, making the ball role from side to side in your palm, Figure 4-8. This works best with a small ball for indoors. For outdoors, you already have the ball that you are bouncing so just use that. Notice how your gait improves while you shake the ball.

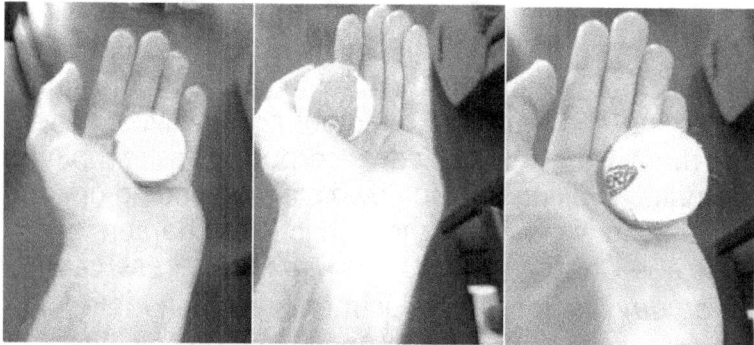

Figure 4- 8

Of course, it does not have to be a ball that is shaken but a ball is a good choice.

Passing a ball back and forth between ha

Simply pass a ball back and forth between your hands and the neuro message pathway opens and the messages get to your legs more clearly, Figure 4-9. You will walk a little smoother. Any small object works such as a rock or stick. If you use the bouce the ball strategy then you can take a break from bouncing and just pass the ball back and forth while you walk.

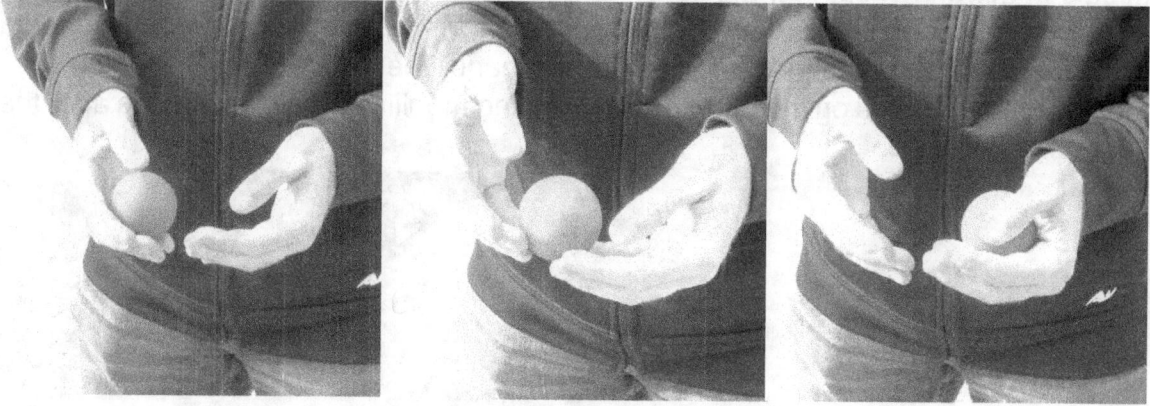

Figure 4- 9

Each technique with a ball should help improve your gait. Alternate techniques to rest your arms muscles. You may find you prefer certain techniques.

Pressure on the palm of your hand

This is perhaps the most important technique because as long as both hands are free you can do it anywhere. It is the answer to increasing speed of movement and to initiating movement. It is the foundation of many techniques.

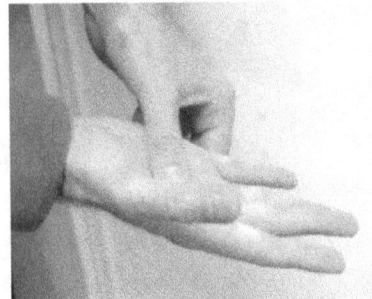

Figure 4- 10

Press on the palm of one hand, Figure 4-10, and within a few steps you'll notice your gait changes and begin to walk more smoothly and heel to toe. Both the sense of feeling and muscle movement are involved. There are many ways that you can incorporate this technique, including walking up and down stairs, standing up, and unfreezing which I'll go through in later chapters. Also see Chapter 13 - *How to use Palm Pressure in Everyday Life* for little adjustments you can make to your habits to take advantage holding objects while you walk.

Another way to make use of palm pressure is to create the pressure with one hand, Figure 4-11. To do this press the fingers of the same hand into your palm. This is useful if only one hand is available. If walking and your arms are shaking it make be more comfortable to use the other method where you press the thumb of one hand into the palm of the other hand. This allows you to better control your tremor.

Figure 4- 11

Snapping fingers

This means you snap the fingers of one hand or both hands by pressing thumb to middle finger and snapping, Figure 4-12. Sometimes if in public simply rubbing fingers together in snapping motion is good enough without making a snapping sound. This helps with leg motion. If done with one hand, the effect of making messages more clear also works on the other arm increasing mobility for a few seconds after the finger snapping.

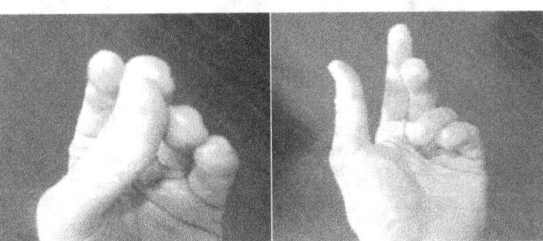

Figure 4- 12

Press index finger into your thumb

Press index finger into thumb as though you will snap your index finger forward but do not actually snap it, Figure 4-13. Each hand should do this as you walk and swing arms. This is the finger flick without the flick.

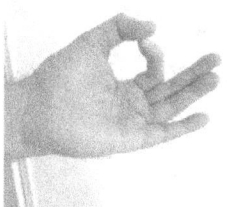

Figure 4- 13

Opening and closing hand(s)

This means you open and close hands continuously, Figure 4-14, either at the same time or by alternating. If both hands are free you can do this. Like the finger snap technique, opening and closing one hand will also allow temporary smoother motion in the other hand as well as better motion in your legs.

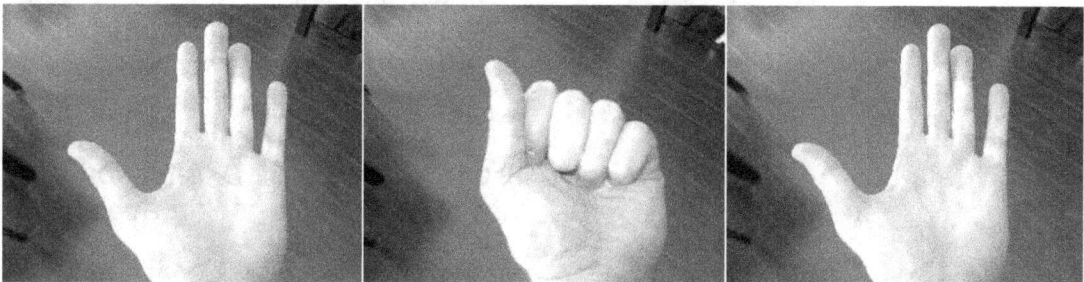

Figure 4- 14

Touch thumb to each finger one after another

Repeatedly touch thumb to a finger, beginning with your index finger, Figure 4-15, and work your way to your pinkie before starting again. Do with one hand our both hands simultaneously. The activeness of your hand and fingers seems to open the neural pathway more, allowing messages to the legs to be received better and improving leg movement while you keep touching you thumb to fingers.

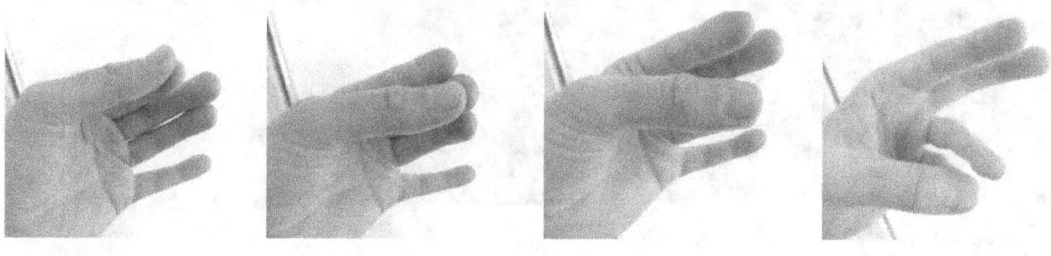

Figure 4- 15

Flick Fingers

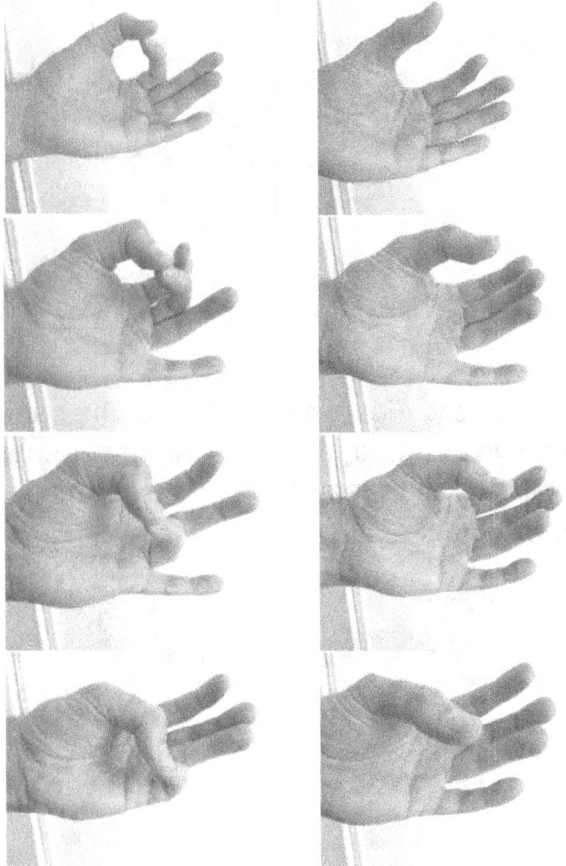

Figure 4- 16

Flicking your fingers provides another option for creating a way to temporarily widen the neural message pathway. It is just another option for a technique so that you have variety and can switch between techniques to give your hand(s) a rest from using another technique.

To do the finger flick technique, begin with your index finger being bent to a position where the fingertip is just below your thumb. Imagine that there is something small, like a rock on your thumb and flick your index finger forward as though you are trying to send the rock flying at something, Figure 4-16. Continue flicking each finger one after the other. After you flick with your pinkie, begin again with your index finger.

Clap Hands

Clapping your hands activates senses of muscles and touch, Figure 4-17.

Figure 4- 17

Carry a small rock in each hand

Rocks in the hands while walking is a great technique to look as normal as possible when out in the streets or on a bike path, for example. Your legs will move better, more heel-to-toe, and you will move faster. You'll also have a better arm swinging motion.

Pick up two small rocks, one to cup loosely in each hand, Figure 4-18. Merely holding the rock seems to open the neural pathway that allows clearer messages to the legs for faster, smoother walking. Holding the rocks keeps the hands and arms busy with nerve messages, which not only allows the arms to swing better but allows messages to transfer to the legs. Your walking strides become smoother, a little longer, and heel-to-toe.

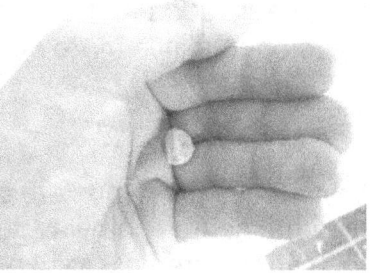

Figure 4- 18

When I'm out I always keep a couple small rocks in my pocket in case I need help with better movement at a mall or someplace I can't immediately find rocks or something else small. Anything small will do such as coins, for example.

Chewing gum

Figure 4- 19

Yes, chewing gum does help, Figure 4-19. Just like how the techniques with the hands will open neuro pathways wider for other messages to travel more clearly through our Parkinson's bodies, chewing gum also temporarily widens the neural pathway by sending messages to the jaw to open and close on the gum. It is most effective you chew. If you use sugarless gum you keep your teeth cleaner at the same time.

Chew very fast for a little more impact. For example, do several quick chews just before doing action that is slowed by Parkinson's such as a boxing punch. Continuous chewing helps ongoing slowed action such as walking uphill. Chewing can be valuable when done in combination with other techniques.

Smile

Smiling, Figure 4-20, uses the muscles of the mouth which triggers the muscle senses and opens the neural pathway to more traffic. Movement in the slowed area will improve for the duration of the time you smile. This makes me think that watching good stand-up comedians and funny shows that cause us much laughter is good for Parkinson's as it will keep our neural pathways open a little more during the viewing.

Figure 4- 20

Sometimes the hands are busy. We can apply the same theory of temporarily widening the neural pathway to the legs and feet. Just keep them busy. Most leg techniques are best to use from a seated position, or if standing if you are leaning on something. They can be combined with chewing gum or pretending to chew gum.

Lift your foot

With your foot flat on the ground, raise the ball of your for foot off the ground, Figure 4-21, and repeat. The movement that you are trying to do should improve while you do this continuously.

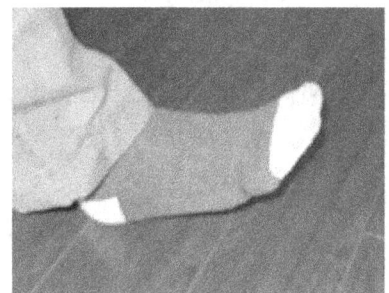

Figure 4- 21

Kick your leg out and back

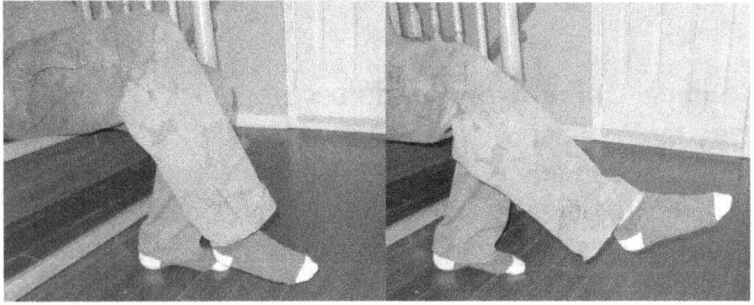

Figure 4- 22

For this technique with your foot off the ground, kick your leg out and swing it back, Figure 4-22. Do this continuously to keep benefitting from widening the neural pathway. There are not many applications for this technique.

Rotate your ankle

Rotate your ankle in circles continuously, Figure 4-23.

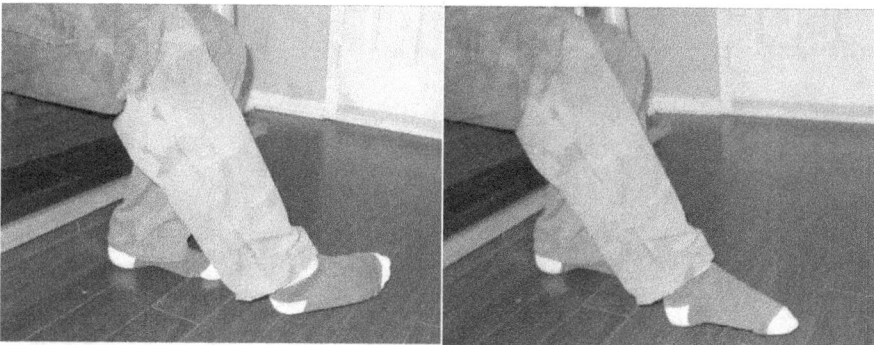

Figure 4- 23

Snap your toes

Snap your big toe with your long toe, Figure 4-24. Keep repeating for maximum benefit.

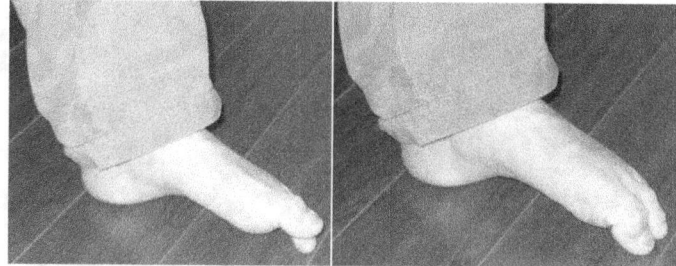

Figure 4- 24

Scrunch your toes

For this technique scrunch your toes, Figure 4-25. You do this by curling your toes and holding them in that position. There are more applications than you would think this technique.

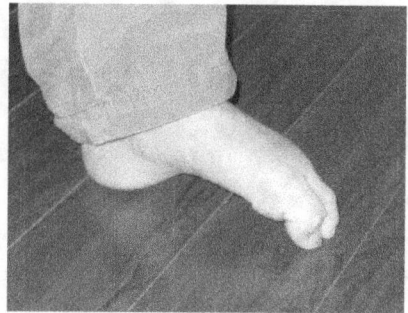

Figure 4- 25

There are a variety of techniques to choose from depending on the situation. You will discover the ones best suited to your needs. Remember you can combine or alternate as needed. The point is to benefit from the way these techniques temporarily widen the neural pathway, allowing you to speed up a motion, better initiate a motion, and add a little missing strength to a motion that you are having difficulty with.

Chapter 5

Improve Slow, Shuffling Gait – Outdoors

Once the stage of Parkinson's is reached where walking begins to slow, our steps become shorter, we shuffle our feet, and over time we walk toe-to-heel rather than heel landing first with the forward leg.

For me this was gradual. I walked well for many years, then at the end of my workday while walking 3km home, partway into my walk and on a slight incline section of my route home, I noticed that I slowed down. If I concentrated on it or had adrenaline kick in if I was late, I could pick up speed, but month by month slow walking at the end of the day became the norm. It was also the time when my medication had run its course. This was only a slow walk; I wasn't shuffling my feet and I was still walking heel-to-toe.

Over time leading up until now, my walking slowed more and during off med times I shuffled with toe-to-heel steps more and more each day. It becomes unpredictable when I'll be affected.

I began experimenting. Years earlier when my main concern was swinging my left arm, I noticed that if I held something in my left hand, keeping that hand busy, I was able to swing that arm better and smoother. I would pick up a twig or a rock or and hold it while I walked. Something I'd carry my keys or a coin in that hand.

As my walking steps became affected and slowed I began experimenting with keeping my hands busy by bouncing a ball, which immediately affected my strides positively. I tried tossing and catching the ball, passing it back and forth between hands. Both worked. It was like the neuro message pathway widened for sending messages and this benefited my neuro walking messages to my legs. But just for a few seconds, meaning the action must be continuous. Gradually I realized new techniques and one key technique is holding the palm upwards and pressing down slightly into that palm with the thumb of the other hand. Somehow engaging the hand with pressure on the palm in this way immediately allows clearer walking signals to get through to the legs. Your stride suddenly improves.

I then thought what else could I actively do while walking? The answer was chew gum. Chewing gum is an active thing to do. You send messages to you jaw to open and close. This also affected my stride in a positive way. On inclines I consciously chewed faster for better results. The great thing about chewing gum is that you can combine it with the hand techniques or alternate between chewing and a hand technique. Hand techniques can also easily be alternated while out for a long walk.

You will have to work harder, sort of multitasking while walking, consciously keeping busy but you will also walk smoother and feel more confident in public. I also think that the more often we walk smoothly the better for us. If I guessed I would say that walking improves by 30% to 40% from whatever point of shuffling you currently are at, at that part of the day. It is significant enough to get me outside more.

When possible wear a jacket so u can carry gum, a ball, and maybe two small rocks so that you can combine and alternate between techniques.

Strategy when walking and using tricks

Using the sensory tricks will enable you to increase stride speed and stride length. My recommendation is that you focus on taking longer heel-to-toe steps, Figure 5-1, in a kind of sauntering manner. Trying to walk too fast can cause you to step toe-to-heel, Figure 5-2, though faster than without the sensory tricks. Longer strides alone will have you advancing more quickly and looks nicer.

When using Neural Pathway Techniques take longer strides rather than faster (shorter) strides.

Figure 5- 1

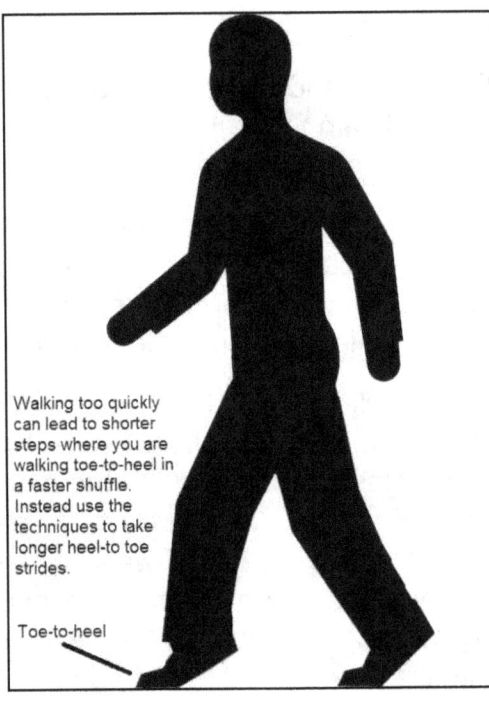

Walking too quickly can lead to shorter steps where you are walking toe-to-heel in a faster shuffle. Instead use the techniques to take longer heel-to toe strides.

Toe-to-heel

Figure 5- 2

Now I'll talk about specific situations and which techniques will help.

Walking outdoors - How to speed up

As Parkinson's progresses walking outdoors to a store, for an errand, or just for exercise becomes something we PD sufferers have to plan, sometimes delay, sometimes cancel, and sometimes dread. We become self-conscious about being out in public and might worry about walking too far while meds are still on and then suddenly really slowing down. But fresh air and exercise is good for us and I hope that these techniques will give you the confidence and drive to continue walking outdoors.

Bounce a racquetball ball or a tennis ball

You want to continuously bounce the ball while walking, taking breaks now and then until your walk returns to a slower pace or until your arms are rested enough.

Figure 5- 1

Try to bounce the ball on a trajectory that keeps your pace fast enough, Figure 5-1. Occasionally the ball might bounce a little out of reach and this is ok as it automatically makes your instincts kick in and you step faster to get to the ball. Once you get a good rhythm going you will chug along pretty nicely. You'll be walking faster and more heel-to-toe and getting out for longer walks. This is also great for dexterity in the arms and hands too. Senses that trigger the widening of the neural pathway: Touch, muscle movement of hands, visual ball target

Using a racquetball ball instead of a tennis ball as there is more bounce to the ball and this allows you to catch the ball palm up because the ball is more likely to bounce higher than your hand so you can catch it on its way back down, Figure 5-2. Palm upwards carrying things is a method that allows nerve message pathways to open a bit so it is a positive to catch the ball this way. A vary back and forth with how high I bounce the ball throughout my walk.

Figure 5- 2

Figure 5- 3

If on a non-busy street you can try flicking the ball up nine or ten feet and a little in front of you as you walk, Figure 5-3. This at times makes you automatically speed up, quickening and lengthening your stride. It's a great feeling when you see how urgency can make you move faster.

Pass a ball back and forth between hands

As you walk, keep your hands busy by passing a ball, or any object, back and forth, Figure 5-4. It attracts less attention than bouncing the ball does and can easily be used as a technique along with bouncing the ball so that you can bounce the ball for a few minutes then switch to passing the ball back and forth in a way that that is little interruption to your stride. This also doubles as an exercise that builds dexterity in your hands.

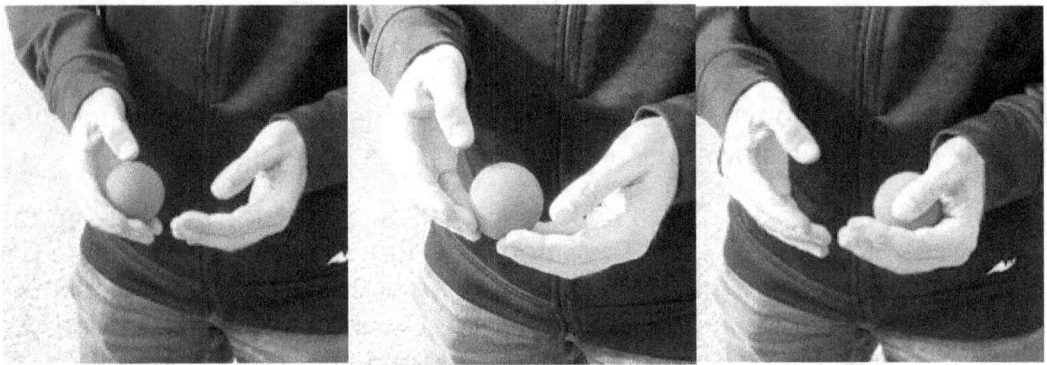

Figure 5- 4

Toss a ball

Tossing a ball is more of an indoor technique but can be done outdoors as well to changeup techniques as you go for a stroll, Figure 5-5. Simply toss the ball a few inches and catch it while you walk. I believe it works well because the palm is facing upwards. The fingers are active too as your hand opens and closes around the ball, increasing messages that temporarily widen the neural pathway. You can toss with one hand or back and forth between hands.

Figure 5- 5

Laser Pointer

The laser pointer when aimed at the ground in front of you, creates a target for your eyes to automatically be aware of. The visual sense is stimulated.

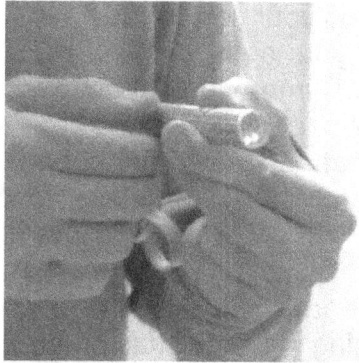

Use a laser pointing device and hold it in your hand while walking, Figure 5-6. You can just hold it at your side not swinging that arm or use while swinging your arm. The key is to be aware of the light on the ground and stimulate your vision with the laser target. For sunny or bright days, use the green laser as it is the most visible for those conditions. For evening or less bright days use the red laser. The same idea applies to using a flashlight. It also stimulation our vision with a target to be aware of

Figure 5- 6

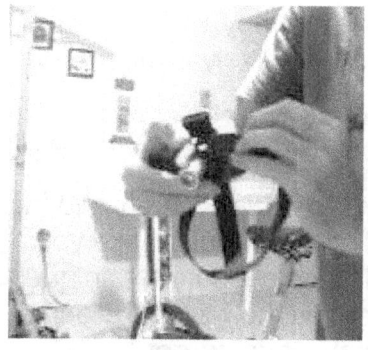

If walking in the evening or when a red laser light is visible on the ground when pointed a few feet in front of you, you can benefit from a headlamp that also has a red laser too, Figure 5-7. It keeps your hands free which is great for when you walk your dog, for example. The flashlight option helps if you have to scoop too. Aim the laser so that the laser light shines the target a comfortable distance on the ground in front of you, maybe seven to fifteen feet. Note that the flashlight of the headlamp works to trigger the visual sense so that if you walk in a dark place and need to use the light, you can switch off the laser and still benefit.

Figure 5- 7

Some ways to use natural visual targets when out walking are listed in Table 5-1.

Table 5- 1

Dog Walking

While you walk your dog, whether on a leash or not, focus on your dog as the pavement moves beneath him, making him a visual target for your senses. Your gait should improve a bit while you do this. Remember, you will improvement but this does not mean you'll walk like you used. Keep expectations to improvement. Trying to keep pace with your dog or not let your dog get to far ahead may also happen by instinct.

Moving Shadow

Take advantage of seeing your shadow as it advances ahead of you. You'll notice an improvement in your gait. Any moving shadow advancing in front of has potential to help.

Walking beside Someone

When going for a walk with someone focus on that person as they walk and use this as a visual target to widen the neural pathways and effectively allow better walking messages to get through to your legs.

In a crowd of walking people

If you are in a crowd of people walking the same direction as you, focus on one in front of you and notice how the ground passes as they walk. This should make your gait smoother. Switch visual targets as necessary. While this sensory technique will likely increase your speed but might not make you as fast as the walkers around you so as the person in front of you changes, you can switch visual target.

Listen to Music with a Strong Beat

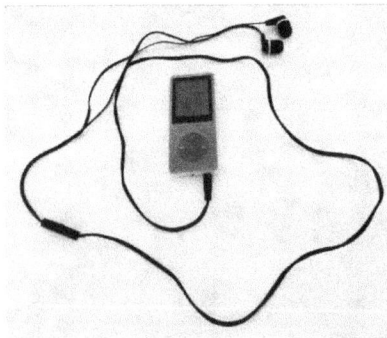

Figure 5- 8

Listen to music on your MP3 or similar device, Figure 5-8. It should have a strong, consistent beat such as in Electronica and House music. The beat will stimulate the auditory sense with a rhythm that will improve messages to your legs. Experiment with and without the music on. This is a convenient technique for crowded places such as malls and airports. You don't need to step to the beat, the beat will however encourage better balance and better strides through the auditory sensory stimulation. Whether u use and MP3 player and earbuds, carry a small player with a speaker, or practice walking near your laptop as it plays, the right music will help you move a little better.

Palm pressure

Pressing the thumb into the upwards facing palm of the other hand is a great technique for short distances and possibly the most important technique, Figure 5-5.

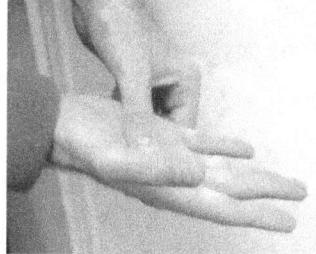

Pushing down on the palm makes the palm hand work in a way that seems to open the neural message pathway for maximum benefit because the palm arm/hand is forced into constant effort. Note that you only need apply a little pressure. This is helpful if you bring the garbage to the curb, for example. Then you must walk back and likely have nothing to hold like a ball or a rock. Use this technique if you are having slow, shuffling moments to change your gait to more of a heel-to-toe stride and walk more smoothly back to the house.

Figure 5- 5

Rub fingers together in finger-snap motion

With hands swinging, rub thumbs on middle finger and index finger repeatedly in a snap of fingers motion or even actually quietly snap fingers, Figure 5-6. This triggers muscle movement senses that will improve neural pathway messages to flow better to your legs as you take each step. To be more discreet put your hands in your jacket pockets, Figure 5-7, and either rub or quietly snap fingers.

Figure 5- 7   Figure 5- 6

Flick fingers as you swing arms

Press index finger against your thumb and then snap your index finger forward, Figure 5-12. Repeat with index finger or continue down the line of fingers. Both hands should do this as you walk and swing arms, though one hand only works as well. You should notice a change to your gait within a few steps.

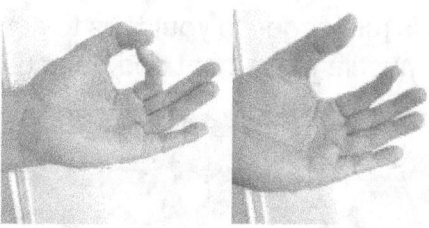

Figure 5- 12

Press index finger into your thumb

Press index finger into thumb as though you will snap your index finger forward but do not actually snap it, Figure 5-13. Continue pressing your index finger into your thumb for continuous sensory benefit. Each hand should do this as you walk and swing arms. This is the finger flick without the flick.

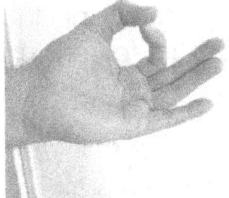

Figure 5- 13

Carry a small rock in each hand

For this technique you need to find two small rocks. You can do like me and always use the same two rocks or you can just pick up a couple rocks from the ground every time.

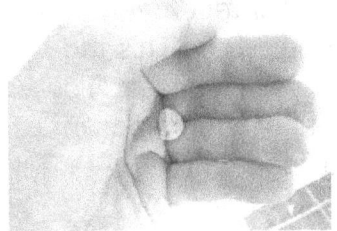

Figure 5- 14

Place one rock in the palm of each hand and sort of cup the rocks while you swing your arms, Figure 5-14. You'll notice that by doing this, your arms will swing more smoothly. You will also be stepping quicker, longer, smoother, and heel-to-toe. The nice part of this technique is that there really is nothing obvious that you are doing while you walk. People won't realize you are holding rocks in your hands. To rest your arms from swinging, if you are wearing a jacket, you can put your hands in the jacket pockets. You want to have the palms facing upwards, still holding the rocks, and not quite resting your hands in the pockets. You want to be still actively holding the rocks.

Open and close hands

This is good for short distances. As you swing your arms open and close your hands which effectively helps temporarily widen the neural pathway and subsequently allows messages to your legs to get through more clearly, Figure 5-15. This makes use of muscle movement and palm pressure.

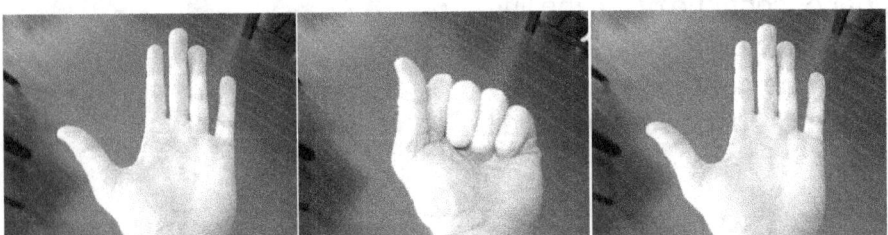

Figure 5- 15

Touch thumb to each finger

Similar to opening and closing your hands, actively touching your thumb to each finger one after another while you walk will improve your gait.

Chewing gum while walking

Chewing gum works, Figure 5-16. If you test it you will experience some benefit. When more impact is needed chew rapidly and you will see your walking pace speed up and your gait improve within a few steps for as long as you continue actively chew quickly. You can fake it if you don't have gum, pretending to chew. Chew gum can be done while using one of the other techniques.

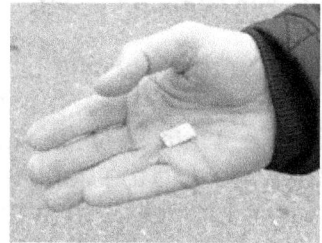

Figure 5- 16

<u>The Walking Warmup</u>

The warmup that I find loosens my legs up for any walk no matter which technique I plan to mainly use, is to flip the ball high and ahead of me so that it will have many bounces before I catch up to it and then, once I am close enough, I catch the ball with my palm upwards. The urgency of getting to the ball, while it still bounces high enough to not have to bend down, just automatically makes me walk faster. Do this for a minute or two, as a warmup, and then switch to the technique you wish to do whether it involves using a ball or not.

Another alternative is to do a hundred foot jog, loosening the legs and arms, then settling into walking. I use this method sometimes when my walking pace is slowing, to jumpstart my speed again. It is easier to run than walk for people with Parkinson's.

<u>Walking Uphill - How to speed up</u>

Walking uphill can present challenges. If you are shuffling and walking more toe-to-heal where you seem to be taking each step more on the ball of or front part of your feet, it is harder to gain momentum as each step is encountering a higher level of ground, Figure 5-11.

Basically each technique for walking outdoors can be applied to walking uphill. You'll find what works best for you.

Figure 5- 17

The nice part about chewing gum is that you can chew while using a hand technique or alternate between chewing and the hand technique. If you carry rocks and the ball, you can alternate many techniques while also chewing gum. I find this to be the best strategy as you can rest different muscles involved in helping in the techniques.

Nordic Walking Poles

Use Nordic Walking poles and go for long walks, Figure 5-18. The arms are kept busy and it keeps them strong. Your arms also get to move allowing them to maintain flexibility. The fact that you grip the handle keeps your hands and arms busy and this in turn leads to smoother movement. If you have balance issues the poles help to stabilize you. Another bonus is that the act of Nordic Walking positions the head upwards which is good for our posture as Parkinson's forces our heads forward over time. It is similar to cross-country skiing in terms of arm motion and muscles use.

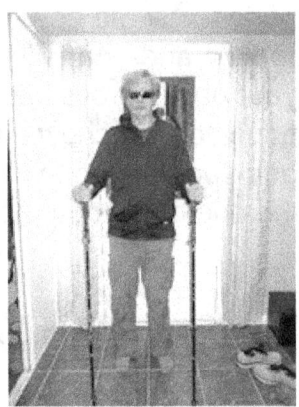

Figure 5- 18

Foldable Walking Pole

You can purchase a foldable Nordic walking pole(s), Figure 5-19 online at websites such as Amazon.

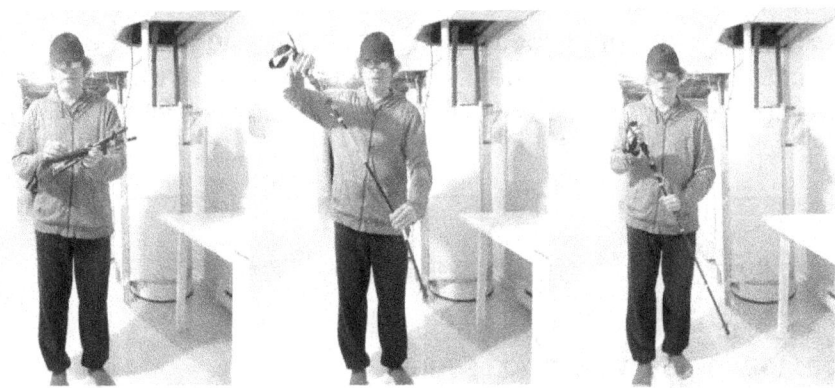

Figure 5- 19

Using One Nordic Walking Pole allows for many techniques

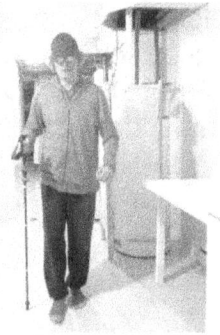

I often use one Nordic walking pole, Figure 5-20, instead of two. The reason is that you can really make use of sensory stimulation techniques to widen the neural pathways. You can walk using the pole as a pole for walking and switch it up using the pole techniques that I will list below.

Figure 5- 20

Neural Pathway Tricks with a Nordic Walking Pole

You can make use of a walking pole to help you walk smoother, more heel-to-toe, with longer strides, and even faster but not in the way you might think. A walking pole is easy to take with you and looks natural enough that you might feel comfortable being seen with it. You can test the walking pole technique with a broom handle or something similar.

Grip the walking pole in both hands and hold palm upwards

By holding the walking pole in a palm upwards motion, Figure 5-21, the senses of the palms trigger stimulation of the neural pathways, allowing messages to the legs to be a little stronger. The muscles of the arms also contribute and actually putting pressure on the pole as though trying to bend it, but not enough to bend it, may help more. Focus on longer heel-to-toe strides rather than faster strides.

Figure 5- 21

Push and pull the walking pole

Hold the walking pole palms downwards with arms bent at the elbows so that your forearms are perpendicular to the ground. Start with the pole pulled towards your stomach. Push the pole away from your body as you are stepping, Figure 5-22. Once pushed away about 1 to 1.5 feet, pull the walking back to your stomach as you continue. Keep repeating as you walk and your gait should improve. It is not necessary to synchronize the push and pull with each stride. I do find that doing a slower push/pull allows me to control having a longer stride and a longer stride more easily means heel-to-toe gait.

Figure 5- 22

Hold the walking pole palms downwards, pushing gently towards your abdomen

Figure 5- 23

Another technique with the Nordic Walking Pole is to hold it parallel to the ground, gripping it palms facing downwards at midsection level while gently pushing the center of the pole on your abdomen, Figure 5-23. Do this while walking and you should notice an increase in control of your strides. You can step faster but I suggest that you focus on longer heel-to-toe steps. The longer stride will advance you faster and smoothly. An alternative is to hold the pole an inch or two in front of your abdomen and put pressure on it as though you are attempting to bend the pole, but not enough to actually bend it. You should see and feel a change in your stride. It is very easy to do either technique and alternate every few minutes if you like.

Remember, you can alternate walking pole technique as you walk. I always do.

Walk pushing alternating pushing each end of the walking pole forward

For this technique you begin by gripping the pole with both hands palms facing downwards and just an inch or two in front of your abdomen. As you step forward with your right leg, push the pole forward a few inches with your left arm while pulling back the same amount with you right arm, Figure 5-24. Then as you step forward with your left

leg, push the pole forward with the opposite arm, the right arm, while simultaneously pulling back with your left arm. Basically whichever leg is advancing, the opposite arm should be pushing forward. Once you have a rhythm going it gets easier. Your stride should improve.

Figure 5- 24

Grip the walking pole in one hand as you swing your arm

Grip the pole, near the center, in one hand and concentrate on swinging both arms, Left arm forward when right leg goes forward and right arm forward as your left leg advances, Figure 5-25. Alternate the hand that holds the pole as needed. Using this technique you get to practice swinging your arms and to exercise your arms with a swinging motion, keeping the muscles used. Gripping the pole activates palm pressure (the sense of touch) as well as muscle movement sense.

Figure 5- 25

Techniques to hold the walking pole under your armpit

Figure 5-26, shows three techniques to hold the walking pole under your armpit. In the first photo you hold the handle tip in the palm of your right arm while the shaft of the pole rests under your armpit. Your left arm goes under the pole with your left hand resting on your right arm. The second photo shows a similar technique, the difference being that your left hand grips the pole near its center. Method three in the third photo involves using only the right arm and armpit to hold the pole. Keeping the arms busy while simultaneously stimulating the neural pathway with sensory tricks allows the focus to be on leg movement and your walking should become smoother.

Figure 5- 26

Hold the Nordic Walking Pole as though it is a spear

For this technique hold the pole as though you are brandishing a spear or a weapon. Think of the grip used on a baseball bat only with your hands further apart. Carry the pole like this as you walk. Your gait should improve. Again, focus on taking longer strides instead of quicker steps. Three ways I like to hold the walking pole in a spear-like fashion are shown in Figure 5-27. Point forward is shown in image 3. When I hold the walking pole in a way that keeps both hands busy so that there is no arm swing, it allow my focus to be concentrated on my legs and their gait. The neural pathway benefit of gripping the walking pole also gets focused on my legs.

Figure 5- 27

Swing the walking pole around your body

As you walk, swing the pole around the front of your body. Start holding the pole in both hands, palms facing downwards. As you step forward, let go with your left hand and pull back with your right arm swinging the forward portion of the pole outward, Figure 5-28. While continuing to walk, swing the pole back in front of you, grab hold with your left hand, then release the pole with your right hand. Perform the same pull back action with your left arm, and continue repeating the right and left side action as you walk.

Figure 5- 28

The benefit is that your arms get some exercise while you still get your legs moving more smoothly.

Arm swing - Hold the pole parallel to your forearm

Another technique to work on arm swinging while simultaneously improving your gait is to hold the walking pole near the center, keeping the upper part of the pole parallel to and touching your forearm, Figure 5-29. Holding the pole activates palm pressure and muscle movement which stimulates the neural pathway to improve both your arm swing and your gait. As always, focus on taking longer steps, as opposed to quicker steps. Once you have a good gait going you might try to increase the speed of your steps. With your other hand you might try using another hand technique at the same time.

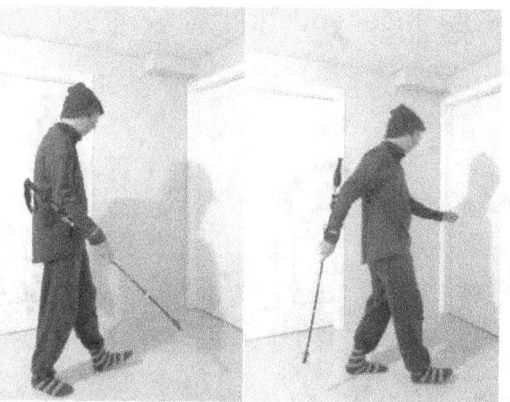

Figure 5- 29

The objective is to get into a smooth rhythm with smoother and longer stride where your arms are doing the best arm swinging motion you can manage. It provides the arm swinging muscles exercise and hopefully helps the muscles remember. Alternate, switching the pole between right and left arm when you feel necessary.

Arm swing - Two poles

If you are using two walking poles, the same technique can be used with a pole in each hand. Hold the poles just a little behind the center of the pole, Figure 5-30. The idea is for the front part of pole to naturally be dipping downwards in the front. As you walk, swing your right arm forward while stepping forward with your left leg, Figure 5-31. At the same time, also swing your left arm backwards. Next as you step forward with your right leg, pull your right arm back, while simultaneously swinging your left arm forward. Easy for a non-PD person but for a mid-stage PWP proper arm swings is something we might have to work at. Holding the poles stimulates the neural pathways allowing better messages to get through to our legs and arms. Holding the poles helps reduce tremors and stiffness in the arms so

Figure 5- 30

it is easier to swing the arms. By holding the poles just behind their centres the poles automatically dip forward a bit; this I have found makes swinging the arms easier.

This technique with two poles allows the arms to more easily get a swinging motion, keeping shoulders looser and arm swing muscles exercise.

Figure 5- 31

Pole under your armpit

Figure 5- 32

Place the tip of the pole handle under your armpit and with the hand of that side grip the pole as far down the pole as you comfortably can, Figure 5-32. You'll be holding the pole kind of the way you would hold a crutch. As you walk swing the pole/arm forward as you would when walking meaning right arm forward when left leg forward, etc. It may take some concentration but you will benefit by working the swinging motion of your arms. Switch arms after a while. You also benefit from improved gait.

Best walking pole strategy

The best strategy is to vary the walking pole techniques. Sometimes holding the walking pole using a two hands on the pole will have the best results, or you just want to rest your arms from swinging for a bit. Sometimes you want to give your arms a swing motion workout. Some days a particular technique just feels right. You may also want to use a Neural Pathway Enhancer in the pole-free hand, see *Chapter 13 - How to use Palm Pressure in Everyday Life*.

Some neat vision tricks

How I learned the following tricks: I lived 2km from the movie theater and was going to be late unless I walked very fast, something stage 3 Parkinson's is not known for. Urgency propelled me. Another completely different scenario had me walking the dog on a snow covered road and I began following boot footprints and trying to match them step for step. My normally shorter steps grew longer and faster as I played this game. Urgency to reach a certain mark and the need have my feet follow footprints propel me to take longer, even faster steps. Longer steps result in heel first foot landings.

Follow footprints on a lightly snow covered road or sidewalk

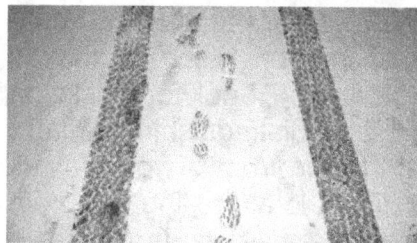

Figure 5- 33

When there is snow on a lightly snow covered road or sidewalk, follow the footprints step for step, Figure 5-33. The bootprints act as a target, much like following a laser pointer's light. Your mind has a goal and a small sense of urgency for your steps to match up with and reach the bootprints with your feet. This works best in light snow as thicker snow is just harder to walk in.

You can still do it in deep snow as long as there are bootprints to follow but you will be slowed by the difficulty of stepping through deeper snow. You will still notice that you step longer and faster when following bootprints. I sometimes follow dog footprints.

Use sidewalk lines as visual targets

Figure 5- 34

While walking on the sidewalk, make a game out of lengthening your stride so as to cover as much of the distance from sidewalk line to sidewalk line, Figure 5-34. For example, your goal might be to take two steps to go the distance from line to line. Possibly you take three steps to cover the distance from line 1, to line 2, to line 3. Basically you are trying to use three strides to cover two sections of the sidewalk. You are using the lines as targets and the number of strides as motivation. Your strides will benefit by lengthening and you might even forget that the goal is to improve your PD walk. At the same time you should feel more control over your arms.

Use natural markers in the street as targets

Figure 5- 35

As you do your walk, tell your brain that you must reach a certain point on the street with some urgency. For example, in Figure 5-35, your goal could be to reach a tree's shadow. Before reaching that shadow, switch your target to the next shadow, then switch to an actual tree or a car parked in a driveway (more precisely a spot on the road that is perpendicular to the tree or car). Look for cracks on the pavement, stones, any distinguishing feature that stands out, as a target in front of you to walk towards. The closer the feature is the more easily effective it is at tricking our brain into sending better walking messages. We have to work a little harder to get a smoother, longer, even faster stride but you will feel less sensitive about being out in public which also means you can get more fresh air and exercise. This method is more difficult than following sidewalk lines as you must continuously look for targets.

Other techniques that can help walking:

Other techniques you can try: smiling, squeeze wooden Neural Pathway Enhancers (see *Chapter 13 - How to use Palm Pressure in Everyday Life)*, secure a washer in the palm of your hand with cord, Figure 5-36.

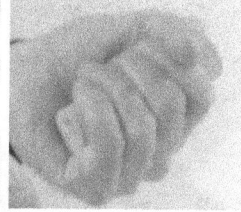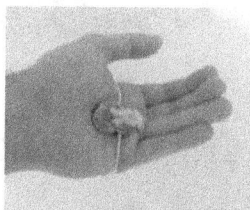

Figure 5-36

Ultimate combination for a fast, smoother outdoor walk
- Two walking poles held just behind the center of the pole
- Walking on the sidewalk, using sidewalk lines as targets
- If at night, with a headed mounted red laser

Chapter 6

Improve Slow, Shuffling Gait – Indoors

The majority of techniques used for walking outdoors can also be used for indoor walking. See *Chapter 5 - Improve Slow, Shuffling Gait - Outdoors.*

Techniques to improve a slow, shuffling gait

Use a laser pointing device.

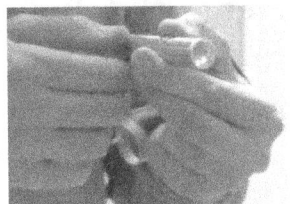

For indoors, use the red laser pointer, Figure 6-1. Either carry it around and use as needed, or you can wear the headlamp laser. I mainly wear the headlamp laser when exercising or walking outdoors at night. For sunny or bright days, use the green laser as it is the most visible for those conditions. For evening or less bright days use the red laser outdoors. Point it at ground around five feet in front of you when walking, though when indoors pointing at walls is useful too.

Figure 6- 1

Music with a strong beat

Figure 6- 2

Listen to music on your MP3, laptop, MP3 radio, etc., Figure 6-2. It should have a strong, consistent beat such as in Electronica and House music. This comes in particularly handy for practicing walking exercises or any exercise. Good for walking around the home doing chores, for example. You do not have to synchronize your steps with the beat though that might naturally happen.

Palm Pressure

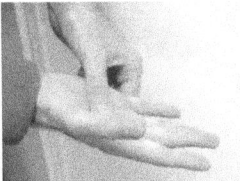

Press the thumb of one hand into the palm of your other hand, Figure 6-3. You will have more balance and walk more smoothly while you press. Chapter 13 - *How to use Palm Pressure in Everyday Life* contains a list of other ways to trigger palm pressure that you might find throughout your home. I use this technique often when indoors or variations of it as will be described in Chapter 13. You are not swinging tour arms when using palm pressure but sometimes you just want to walk without shuffling.

Figure 6- 3

Flick Fingers

 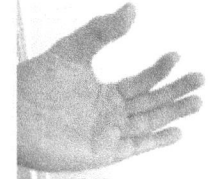

Flick fingers, Figure 6-4, to boost your control as you walk around the house. This is convenient for short distances. You can flick on finger after another or just continuously flick one finger.

Figure 6- 4

Walk holding a short stick in both hands

Walk holding a short stick in both hands, something like a shortened broomstick handle, Figure 6-5. This is an indoor substitute for using a Nordic hiking pole. The stick should be about shoulder width. Two useful grips to employ are palm downwards

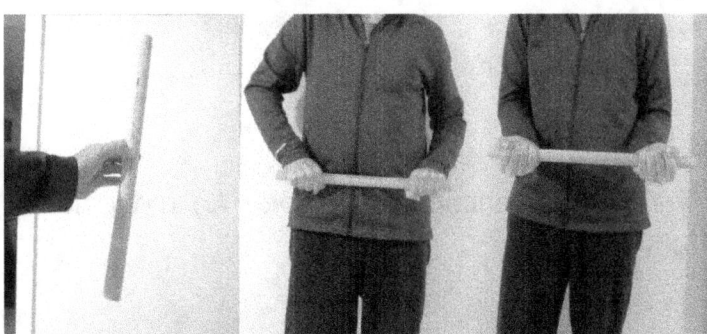

Figure 6- 5

and palm upwards. For either method, it can be effective to put a little pressure on the stick, as though you are trying to bend it. This helps maximize the neural pathway messages to the legs. I normally only use this particular short stick when doing balance exercises. It is a variation of what I do outdoors when walking with a Nordic walking pole.

Use tiles and their lines to reduce your shuffle

Similar to using the sidewalk lines, have your brain tell your legs to step covering two tiles, three lines, with each step, For example. Your visual sense will set the lines up as targets, Figure 6-6. Your stride will instinctively lengthen and become smoother. At the same time you should feel a loosening in your arms and your arm swing.

Figure 6- 6

Walking during the middle of the night bathroom break

If you are like me then your bladder has weakened and you wake often to use the washroom. You just woke and you want the simplest technique to help you walk to the washroom and back to bed.

Palm pressure by pressing the thumb of one hand into the palm of the other is the simplest technique and very effective.

You can also rub thumbs on middle finger and index finger repeatedly in a snap of fingers motion or even quietly snap fingers, do with best hand or both hands

Another is to open and close hands repeatedly either at the same time or by alternating as you walk.

Using the hand held laser pointer is a good option, especially if it has a flashlight option, or use an actual flashlight. Remember the light on the floor acts as a target that improves your walking.

Ultimate combination for indoor walking

- Squeeze a wooden Neural Pathway Enhancer (see *Chapter 13*) in one hand
- Use lines of floor tiles as targets for stepping

Chapter 7

Stairs

Walking up or down stairs

Stairs present an increasing daily challenge the more bradykinesia slows you down and the more off med moments you have. Fortunately my techniques can be applied to help. By applying one of the techniques you will feel a 30 to 40% increase in smoothness and speed as you climb or descend the stairs. Your balance should also improve, making stairs a more safe experience. You will feel more confident and not dread the need to use the stairs.

Ascending stairs

Up-the-stairs - Finger-snap

Use the finger-snap technique to boost your speed, balance, and control when climbing stairs, Figure 7-1. Snap the fingers of one hand as you ascend. If you feel uncertain, use one hand for keeping hold of the railing while the other does the finger-snaps. You should notice your steps are a little easier, a little faster. You can snap quietly, though a consistent sound may be beneficial.

Figure 7- 1

Up-the-stairs - Open-close hand

You can also open and close your hand(s) as you climb up the stairs, Figure 7-2. Your hands can just rest at your sides or be swinging or just be held in front of you, each method will make you move more effectively upwards.
Have someone watch you from the top of the stairs as you climb normally, then with the finger-snap technique and the open-close hand technique. They can verify what you already feel, that you have improved on the stairs using the technique.

Figure 7- 2

Up-the-stairs – Palm pressure

A favorite technique of mine, also can apply to the stairs. It is the palm pressure technique, Figure 7-3. By keeping constant pressure on your palm you get a constant benefit going to the neural pathway and consistent stepping on the stairs. As long as your balance is boosted enough to make using the stairs safe enough without the use of one hand on the railing, the palm pressure technique is a good option.

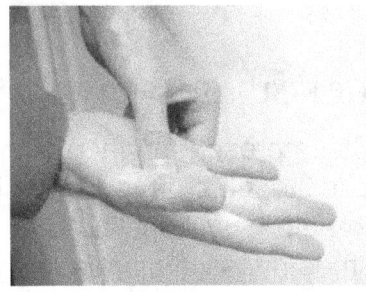

Figure 7- 3

See *Chapter 13 - How to use Palm Pressure in Everyday Life* for more ways to create palm pressure that are convenient when indoors.

Up-the-stairs – Toss-a-ball

Less convenient and more risky because it is on the stairs is the toss-a-ball technique. I mention it to demonstrate it can work too. You can do short upward tosses with the smaller indoor ball while the other hand while holds the railing while going up or down the stairs.

You can also just shake the ball in your hand which is less distracting to your focus of the stairs, Figure 7-4.

Figure 7- 4

Up-the-stairs – Laser pointer

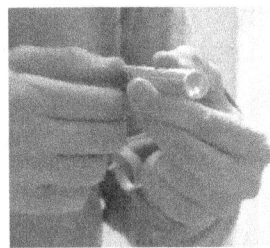

Figure 7- 5

The laser pointer (indoor red laser) can be used to stimulate the vision and boost stair climbing control. Just shine the laser in front of you where you can easily be aware of it as you ascend. The mere fact that your hand is also busy holding the laser pointer device also helps maximize improved stair navigation. Keep in mind that it is maximizing improvement that your current stage of PD, current on/off med level will allow. It does not mean perfection, only improvement and more independence.

Up-the-stairs wine bottle palm pressure

A version of palm pressure is to use an object by carrying it in the palm of your hand. For example, if you go get a bottle of wine, carry it in the palm of your hand and specifically in this case up the stairs, Figure 7-6. Your ascent will improve.

See *CHAPTER 13 - How to use Palm Pressure in Everyday Life* for more examples of palm pressure.

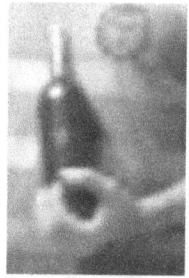

Figure 7- 6

Descending Stairs

Techniques that work for ascending stairs can be used for descending stairs.

In Figure 7-7, I descend the stairs while snapping my fingers.

The techniques help initiation of leg movement, control, speed, and balance, making descending stairs safer.

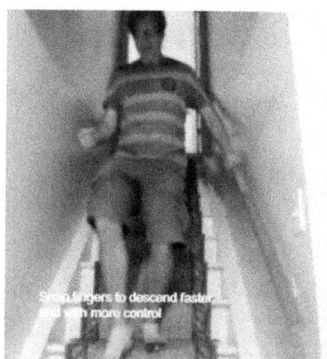

Figure 7-7

Other techniques that can help with stairs:

Other techniques you can try on the stairs: smiling, chewing gum, squeeze a small piece of broom handle, music with a beat, press index finger into thumb and hold pressure, Figure 7-8.

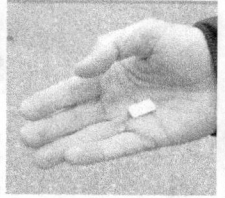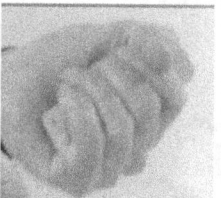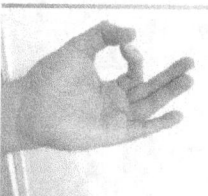

Figure 7-8

Ultimate Combination for stairs
- Palm pressure (see *Chapter* 13 for alternative ways to create palm pressure)
- Smile

Chapter 8

Stand up from a Seated or Laying Position

I take as little Parkinson's medication as can manage due to an effort to keep side-effects as low as possible. So there are times when just standing up can be a slow-motion struggle. Interestingly sometimes when on what should be peak performance of meds, I seem to struggle more. When meds wear off is likely the biggest struggle. Note that a couple hours after the final wearing off I do begin to move relatively well. Basically, there can be numerous times in a day, depending on the day, wear these techniques can be employed. In terms of standing up I prefer the following.

Chair to stand

Chair to stand using finger-snaps

Getting up from a chair can become a slow motion challenge when Parkinson's meds are low or off. Try this, when you are having difficulty getting out of a chair, whether a recliner, bench, stool, or dining room chair, first attempt to stand as you normally would. It can be a struggle that might take a few attempts, Figure 8-1. Possibly your meds are not completely off and you manage to get up on the first try, but a little slowly. Now try again, only this time with one hand continuously snap your fingers as you rise from a seated position to a standing position, Figure 8-2. You can use your other hand to help pushing off the armrest if there is one.

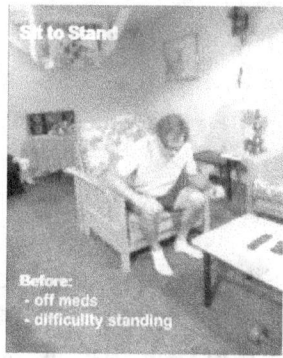

Figure 8- 2

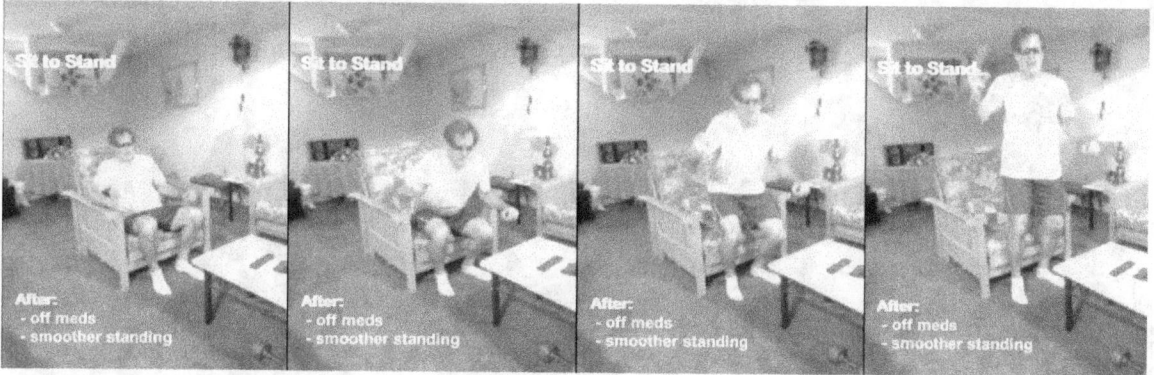

Figure 8- 1

You should experience a noticeable improvement in your ability to stand.

There are other tricks you can do. In the following example, I make use of a small jar of nuts that is on the coffee table.

Chair to stand using a jar of nuts

While still seated, place the jar into the palm of one hand. This activates what I call Palm Pressure. Both the sense of touch and muscle movement stimulate the neural pathways, allowing increased initiation of movement, increased control, and increased speed. The result is that you will rise to a standing position quickly, with less effort, and with better balance. You can use your other hand on an available armrest, or dining table, or the seat itself to help guide you to a standing position, Figure 8-3.

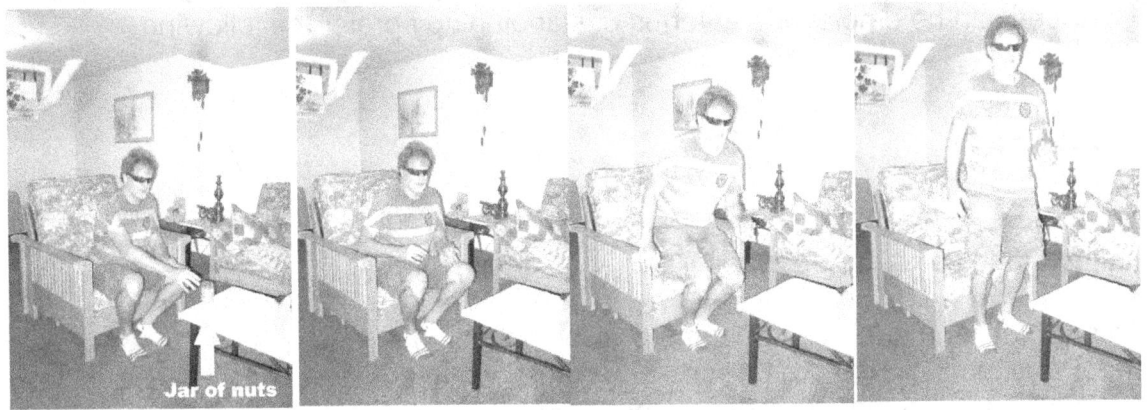

Figure 8- 3

Stool to stand using finger-snaps

Sitting on a stool (no armrests) is a good way to tests the neural pathway sensory tricks. We will use the finger-snap technique to go from sitting on the stool to standing, Figure 8-4. Begin snapping the fingers of one hand as you begin to move to stand and continue snapping until standing straight up.

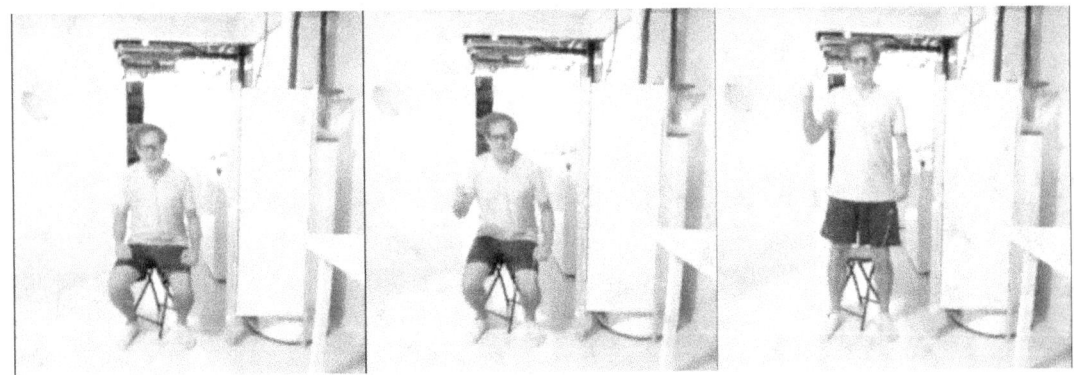

Figure 8- 4

More options for going from sitting in a chair to standing

Getting out of a chair has more options for techniques than getting out of a sofa because of the position that you start in. Like for the sofa, you can shake a ball, rub or snap fingers, and chew or pretend to chew gum. If no one is around you can make a big smile, using mouth muscles to stimulate the neural pathway.

You can also open and close your hands while getting to a standing position or just open/close repeatedly one hand while using the other for leverage. Palm pressure can be used if you feel you can stand without using one of your hands for leverage or balance, Figure 8-8. Press your thumb into the palm of your upwards facing hand until you are standing. Keep pressing if you are going to walk. An alternative, to keep one hand free, is to hold something in your hand with palm facing upwards, like a cup, pencil, paperclip, bottle-cap, ball, or other nearby item.

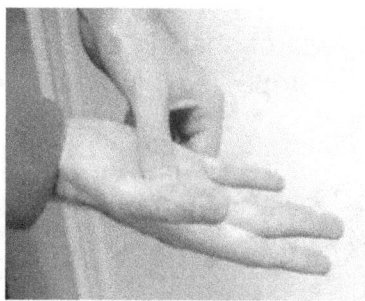

Figure 8- 5

Laying in sofa to standing

A few techniques can be used to help you get from a laying position on the sofa to a standing position when Parkinson's bradykinesia slows down your movements, making getting off the sofa a struggle.

Sofa to standing using a small indoor ball

You can keep a small ball for indoor use and use it for techniques as you walk around your home. If you have the ball handy when you are laying on the sofa, it can be used to speed up your process of standing up. With your best hand or convenient hand, shake the ball from side-to-side rapidly while turning your body in the sofa. Continue shaking the ball while swinging your legs to the floor and moving to a sitting position, then proceed to stand, Figure 8-5.

Figure 8- 6

You must shake the ball the whole time and can use your free hand to help you as needed for leverage. Experiment with this and compare to see how much faster you move.

Sofa to standing finger rub or snap fingers technique

Instead of shaking a ball, this time rub your thumb on your middle finger and index finger repeatedly in a snap of fingers motion or even quietly snap fingers. Do with

your best hand or both hands, Figure 8-6. Whichever method you use to help you move more quickly to get to a standing position the key is to continuously use the technique with one or both hands to get maximum benefit. You'll find what works best for you and possibly you'll prefer another technique. If you happen to have gum in your mouth then you can use it to help by chewing rapidly until you are standing. If you don't happen to have gum, you can pretend to chew rapidly. It will still help.

Figure 8- 7

Laser Pointer

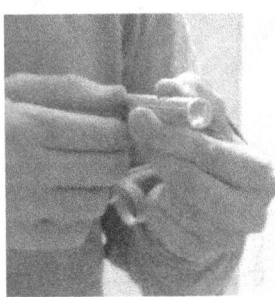

You might be in the habit of carrying the laser pointer with you while indoors, Figure 8-7. If you are struggling to stand you can point the laser where you can easily see it as you go move from a seated position or relaxed position on the sofa to a standing position. It is a good transition to walking to because if it is difficult to stand, you likely will shuffle as you walk and the laser will help reduce shuffling.

Figure 8- 8

Lying in bed to standing

This is similar to getting off the sofa.

Laser pointer - Lying in bed to standing

Make use of the laser pointer to create a target to stimulate the visual sense. While lying in bed, point the laser at the ceiling and look towards the laser spot on the ceiling, Figure 8-9. Begin maneuvering and rolling towards the edge of the bed. As you turn, shifting your feet off the bed point the laser at the wall and continue until you are standing. There should be a noticeable improvement in speed and smoothness.

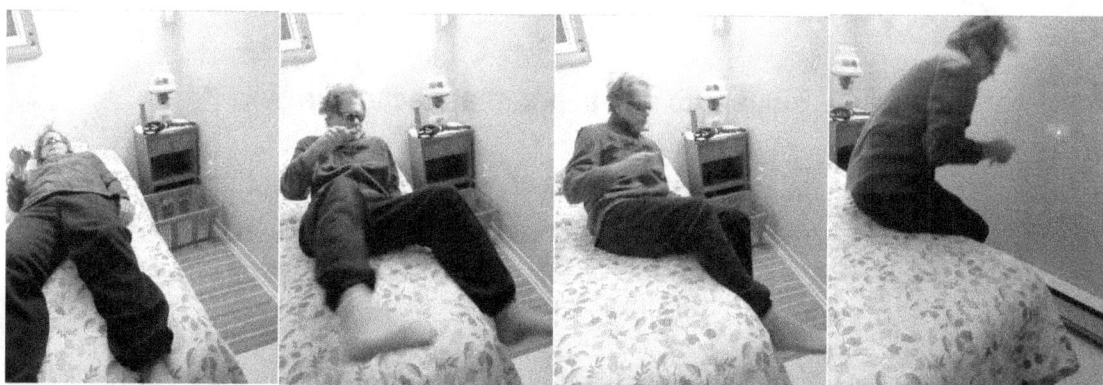

Figure 8- 9

Shake a ball - Lying in bed to standing

Here is another technique that can be used, this one relying on the sense of touch and on muscle movement. With your best hand or convenient hand, shake the ball from side-to-side rapidly while turning your body in the bed, Figure 8-10. Continue shaking the ball while you use the opposite hand to help you turn.

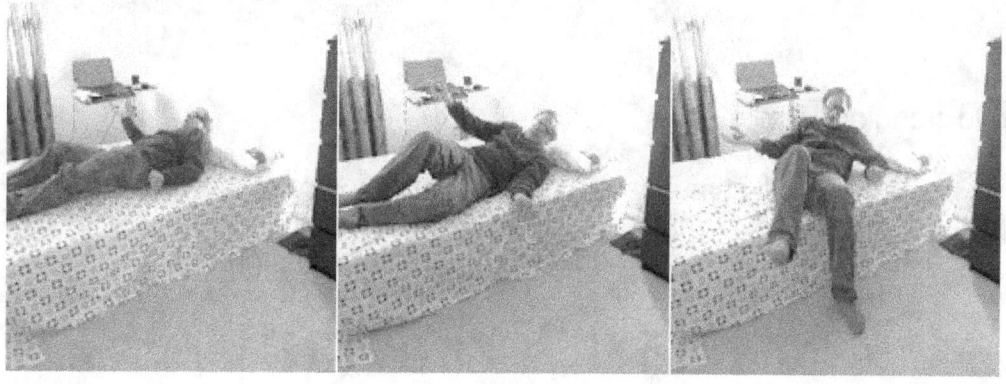

Figure 8- 10

Continue shaking the ball while swinging legs to the floor and moving to a sitting position, Figure 8-11, then proceed to stand at which point the shaking ball technique can be replaced to the toss the ball technique if you like). Note: Most beds are higher than sofas but I like this futon that opens up into a bed because it is firm, more firm than most beds. This makes maneuvering to rollover or get out of bed it easier.

Figure 8- 11

Lying in bed to standing - Finger rub and finger-snap technique

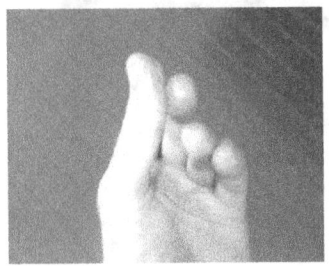

Figure 8- 12

Instead of shaking a ball, this time rub your thumb on your middle finger and index finger repeatedly in a snap of fingers motion or even quietly snap fingers. Do with your best hand or both hands, Figure 8-12.

You may need to use one hand or at times both hands to help you maneuver around one the bed but try to keep one hand rubbing or snapping while you do this.

Lying in bed to standing - Chewing Gum

If you happen to have gum in your mouth then you can use it to help by chewing rapidly until you are standing. If you don't happen to have gum, you can pretend to chew rapidly.

Lying in bed to standing - Smile

Figure 8- 13

Smiling will stimulate the neural pathway through jaw muscle use, Figure 8-13. It might seem silly to do but try it and see an improvement as you exit the bed. This suggest that laughter is good too.

Turning in bed - Techniques

Turning or rolling over in bed can be a frustratingly slow process once bradykinesia develops. A firm bed helps.

In terms of techniques:

- If you have a ball or small object, you can use the shake the ball technique while you change positions.
- You can rub or snap your fingers
- You can use one hand palm pressure by closing your hand and pushing your fingertips into the palm of the same hand.
- You can open and close your hands until your change of position is complete.
- You can pretend to chew gum rapidly as you turn.
- You can smile to stimulate the neural pathway.

From sitting on the floor to standing - Techniques

See techniques for getting up from bed, from a sofa, and from a chair.

Initiating a movement with bad arm

The techniques help temporarily with improving movement in the other arm/hand, improving movement by up to 30 or 40%. Sometimes you may want temporary improvement in your better arm, for example when brushing your hair. Sometimes you might wish to add improvement in initiating movement in your weaker, more affected arm, for example when hitting a punching bag.

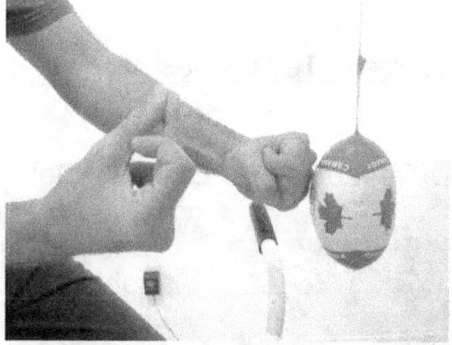

To help with movement in an arm that is having trouble with initiation of movement while punching or boxing, snap the fingers of the other hand just before you want the troubled arm to move, Figure 8-14.

Figure 8- 14

Ultimate Combination for standing

- Finger-snap
- Smile or clench teeth

Chapter 9

How to Unfreeze

As of the time of writing this book, I have had moments where my legs froze. After stopping and standing in place for a few seconds or longer when I tried move my legs, I could not make either move. I could not step forward. I could not turn. I could not make my either leg move no matter how many times my brain tried sending messages to my legs.

For me, at this stage (stage 3), it happens when my meds were low or wearing off. Imagine if the freeze happened in public on stairs, or on an escalator. In the middle of crossing a street would be one of the worst scenarios, Figure 9-1. What you want is to not freeze, of course, but if you do freeze then what you want are methods to unfreeze as quickly as possible.

Figure 9- 1

Freezing is the extreme of initiation of movement issues and so far the most difficult to overcome, with the exception of fatigue which I haven't figured how to deal other than napping. This chapter will present techniques that I use to unfreeze that work for me.

Unfreeze Techniques

Unfreeze using a Laser Pointer

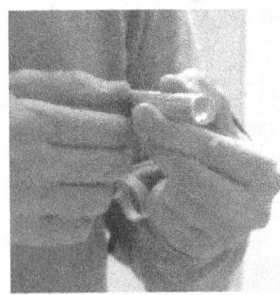

Figure 9- 2

Carry a laser pointer with you in case you freeze. Use a red light laser pointer for indoors, Figure 9-2, and for outdoors during the evening. If you are not actively using it to improve your gait while walking, have it easily accessible. Note that if you are already using it to help with your gait, your legs probably won't freeze up in the first place, at least this has not happened to me.

For outdoors during bright days, you'll want to have a green laser light laser pointer easily accessible as the green light is more powerful and can be easily seen even on bright sunny days.

The assumption that I am making is that it is your legs that freeze but that your arms/hands are unfrozen and mobile enough to reach for the laser pointer. When your legs freeze shine the laser anywhere that is within your line of sight, Figure 9-3. Focus on the dot created by the laser light, though you do not have to look directly at the dot and you should feel control return to your legs and be able to immediately move your legs. You do not have to point the laser pointer yourself, another person can be the one pointing its dot as long as it is visible to you.

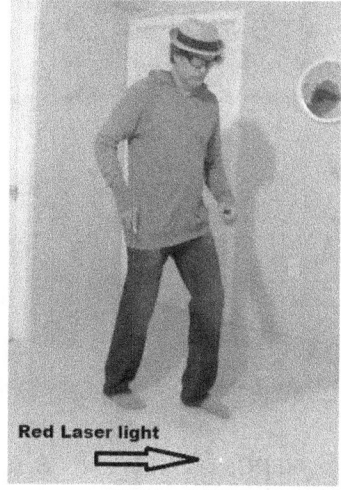

You can also try this using a flashlight instead of a laser pointer, as long as the light is visible enough and distinctive enough. I actually would walk the dog in the evenings with a headlamp shining on the ground about ten feet in front of me, until I bought a headlamp that had the option of using a head-mounted red laser pointer.

Red Laser light

Figure 9- 3

Unfreeze using the palm pressure technique

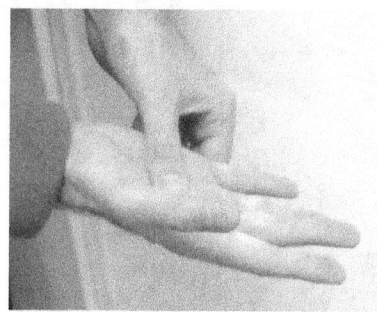

Figure 9- 4

The palm pressure technique might be the second most effective method to unfreeze your legs, after the laser pointer, Figure 9-4. Press the thumb of one hand down on the upward facing palm of the other hand. The palm upwards hand should be resisting the pressure. For me at the stage of PD I'm at, stage 3 PD, this has worked 99% of the time. I wish I could say 100%. But it absolutely has worked and allowed the messages from my brain to my legs that they should move to get through and my legs do then move.

Unfreeze, Palm pressure in general

Let's say you are in the shower and your meds run low. You begin moving slowly. Your tremors increase. Now your legs freeze. What can you do? To unfreeze reach for the top of the shower, Figure 9-5, grab the top, and press downwards so that your palm is pressured. You should be able to move your legs at this point. You can press your palm against the wall as well.

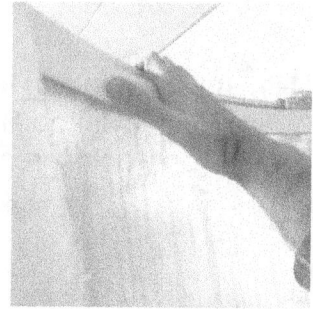

Figure 9- 5

See *Chapter 12 - Personal Care, Bathroom, and* Showers for more bathroom and shower specific advice.

Snap Fingers

You can try snapping your fingers, Figure 9-6. For lesser freezes this technique may help. You can also make a fist and squeeze very tightly.

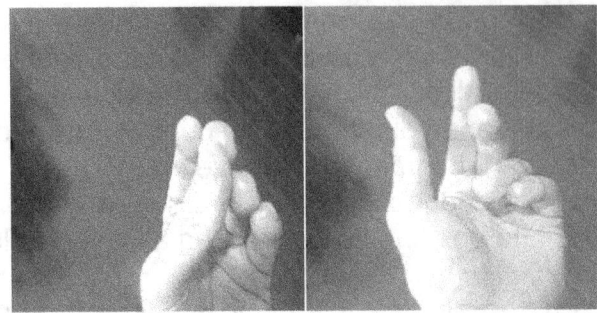

Figure 9- 6

Unfreeze by smiling

Smiling, Figure 9-7, is not as effective at unfreezing your legs as a laser pointer or pressure on the palm while you attempt to move your legs. You can also clench your teeth or pretend to chew gum. The use of the mouth muscles is a sensory technique that allows messages to travel through the neural pathway a little more clearly. These mouth/jaw techniques are good to use in conjunction with other techniques.

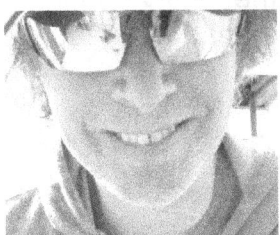

Figure 9- 7

If walking outdoors, Figure 9-8, I make it a practice to already be chewing gum which in itself might help to not freeze in the first place as chewing gum does help temporarily allow messages to travel more clearly through the neural pathway. If you do freeze you and have no gum in your mouth, combine pretending to chew gum or clenching your teeth with one of the other techniques to unfreeze your legs.

Figure 9- 8

Ultimate combination to unfreeze
-   Point laser pointer light at the ground
-   Apply palm pressure to the hand holding the laser pointer

Chapter 10

Clothing and Getting Dressed

With hand dexterity decreasing a feeling of loss of independence might be experienced as you occasionally rely on help. The techniques that follow can help prolong some independence. Here are some strategies for getting dressed when you are having difficulty initiating movement and controlling your arms.

Putting on a t-shirt

On a bad day, something as simple as putting on a t-shirt can frustrating and time consuming. Our hands just won't function properly. Someone might have to help. Instead of asking for help there are techniques you can try.

Putting on a t-shirt – Method 1

One technique is to grab the back collar of the t-shirt in a tight fist, Figure 10-1. Continue gripping tightly and with the other hand grab further down the shirt and scrunching up more of the shirt in that hand. Repeat back and forth until reaching the bottom of the shirt. The t-shirt is now ready to pull on.

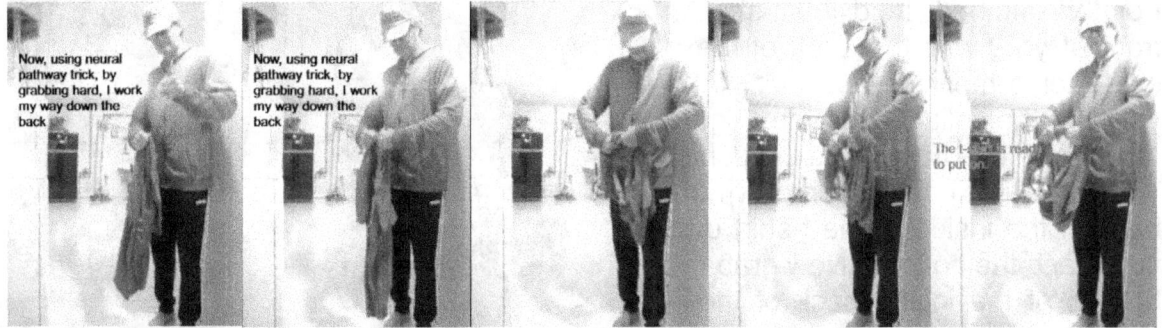

Figure 10- 1

The key is that the hand that is holding the shirt should squeeze tightly. By gripping tightly both the sense of touch and muscle use of the hand help improve movement in the opposite hand as the neural pathways temporarily widen.

Putting on a t-shirt – Method 2

Place the t-shirt unfolded on a table, bed, or any flat surface. Grab the back bottom of the back of the t-shirt in a tight fist. Continue gripping tightly and with the other hand grab further down the shirt. Repeat back and forth until reaching the collar of the shirt. The t-shirt is now ready to pull on.

Figure 10- 2

Putting on a t-shirt – Method 3

For this method, hold the t-shirt in front of you, with the back of the t-shirt facing more or less away from you and the collar at the top, Figure 10-3. With one hand, hold the t-shirt. With the other hand reach down the back of the inside of the t-shirt until you reach the bottom. Now grab the bottom of the inside back of the t-shirt and pull upwards reaching the collar. With that same hand also

Figure 10- 3

grab the collar and everything in between. You are now ready to pull the t-shirt over your head.

Putting on a jacket

Some days Parkinson's makes putting on a jacket a difficult process. By gripping hard at each step of putting on the jacket, you
can stimulate the neural pathways, improving your other arm's
control to do what it needs to do. By gripping tightly, I mean you
hold the jacket in a tight fist with one hand, Figure 10-4, while
maneuvering the other arm. Basically you alternate gripping part
of the jacket as you go. It should feel like a noticeable improvement. Begin by gripping the top right collar of your coat with
your left hand, making a tight fist, Figure 10-5. Slide your right
arm through the sleeve. Next, once your right hand is through,
grip the end of the sleeve with that hand in a tight fist so that the

Figure 10- 4

end is also in your right palm to stimulate palm pressure and muscle movement. Your
left arm should be a bit more controllable so now slide your left arm through the
jacket's left sleeve. Repeat this process as you adjust the jacket on you, using one
hand to hold a part of the jacket tightly.

Figure 10- 5

You can experiment with techniques when getting dressed. See what works best for
you when you button up a shirt, or put on and buckle up a belt, or wrap a scarf around
your neck. Maybe in the winter pulling on or pulling off long johns is difficult. Experiment and see what works best for you.

Putting on shorts

To improve both your balance and arm control when putting on and pulling up shorts, stimulate the neural pathways by grabbing the shorts in a tight fist for the duration of the time that you are putting on the shorts, Figure 10-6. While gripping tightly, my balance is better and I am able to control my arms better and to pull the shorts up the final bit. Note: You may to be cautious and to lean on something while putting on shorts.

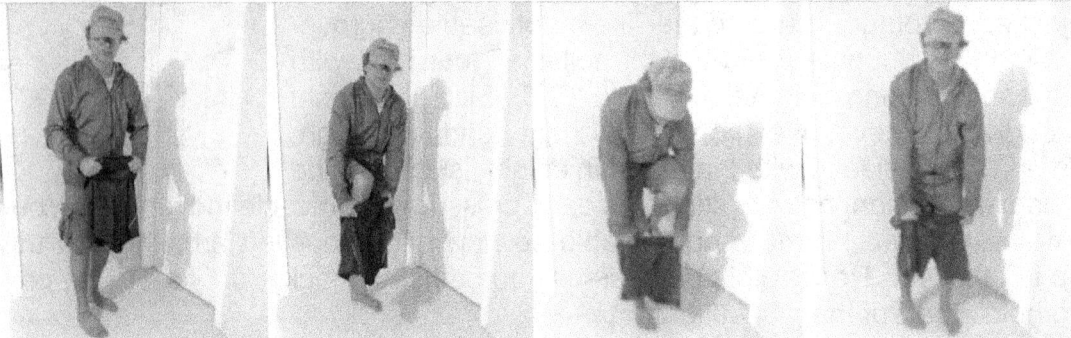

Figure 10- 6

Other dressing techniques:

Other techniques you can try in conjunction with the dressing techniques or on their own while you get dressed are: smiling, chewing gum, scrunching toes, music with a beat, setting a laser pointer down and having it point to a visible location, Figure 10-7.

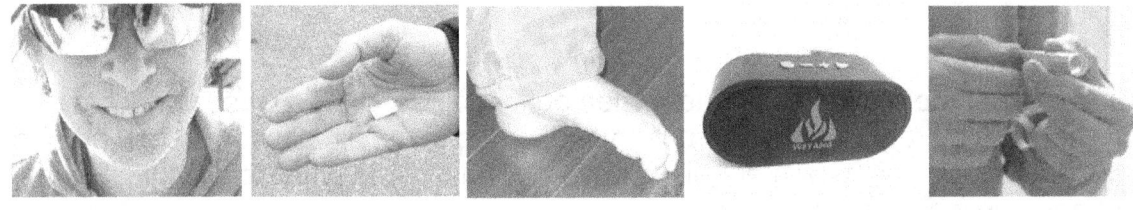

Figure 10- 7

Ultimate combination for getting dressed
-   Grip clothing tightly
-   Smile

Pockets are good

Jacket with pockets

Figure 10- 8

Figure 10- 9

Wear a jacket with pockets whenever you can. With the difficulty that Parkinson's can make it to swing our arms, sometimes I like to rest my arms by putting my hands in pockets. I find it easier and more comfortable to do this with jacket pockets, Figure 10-8, than with pants pockets when walking. You can also keep your hands in your jacket pockets and use some hand techniques like pressing index finger into your thumb or holding a small rock in each palm. Another advantage to pockets is that you can carry those same rocks and/or a ball for bouncing in the pockets, Figure 10-9.

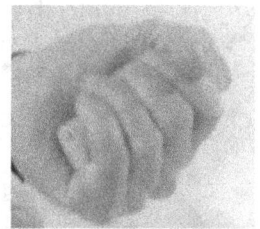

Figure 10- 10

You might also carry a sawed off broom handle in your jacket pocket, ever ready to be tightly gripped, Figure 10-10, and hidden with your hand in your pocket, or actively used/gripped as you swing your arm.

Fanny Pack

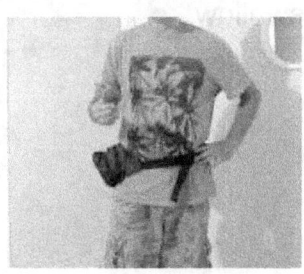

Figure 10- 11

A fanny pack, Figure 10-11, allows you to carry many useful tools to help you walk smoother. You can carry gum, rocks, a ball, the sawed off broom pieces, a laser pointer, the Neural Pathway Enhancers, etc. This allows you to vary the techniques as you walk.

Chapter 11

Swallowing & Miscellaneous Issues

I take vitamins every morning and since reaching mid-stage of Parkinson's getting vitamins or meds stuck in my throat, as well as food stuck in my throat has become an ongoing and uncomfortable concern. Always I keep enough liquid ready in case something gets extra stuck. I eventually tried neural pathway techniques hoping they would affect throat muscles. They did. I also combine them with a swallowing strategy.

Swallowing Technique

I've experimented with a variation of the palm pressure technique and I feel it helps the messages to the throat muscles. But don't be overconfident, but do be careful when applying the following technique and have enough water or coffee ready in case

Figure 11- 1

the vitamin or food does get stuck. For me I find there is an improvement but that still does not make it back to normal, so careful please.

After putting the vitamin in your mouth be sure to pour enough water into your mouth, Figure 11-1.

Then hold the cup in the palm of your hand, Figure 11-2. Now swallow.

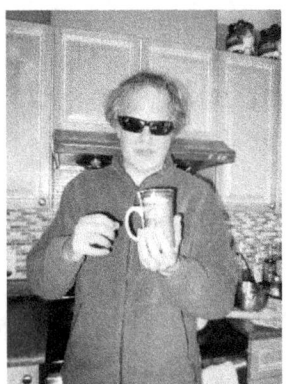

Figure 11- 2

Freeing stuck vitamin/pill/food

If the something gets stuck in your throat when you are swallowing, try this technique. Pour more water into your mouth, hold the cup in the palm of your right hand while using the left hand to reach to the top of the right side of your head and then to pull your head to the left side, which allows another angle for the vitamin to follow down your throat, Figure 11-3. Now swallow. It may be necessary to try the reverse by holding the cup in your left hand this time and using your right hand to pull your head to the right. Repeat as necessary.

Figure 11- 3

I actually have the habit of always using the tilt-head method, combined with the cup-in-the palm technique whenever I swallow vitanims or meds. For food I only use the tilt-head method (and cup-in-the-palm technique) if food has gotten stuck in my throat.

Swallow while using a laser pointer

Either hold a red light laser pointer or wear the head-mounted laser, Figure 11-4, and point the light where it is easily visible while you are swallowing. This uses the sense of sight to temporarily increase message traffic to the throat muscles, making your control of those muscles a little stronger. If the light is visible as a target, you can use a flashlight instead.

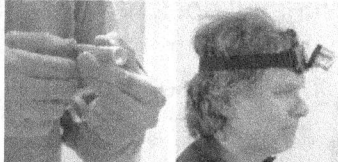

Figure 11- 4

Swallow while holding a sawed off piece of broom

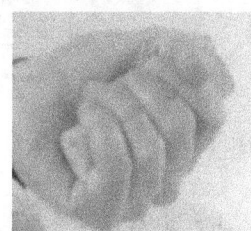

The piece of wood can be made from a broom handle and sawed to a length slightly longer than the width of your closed fist, Figure 11-5. Grip the wood tightly as you swallow, triggering both palm pressure and hand muscle use to open the neural pathway to send a clearer message to the throat muscles for swallowing.

Figure 11- 5

Ultimate combination for swallowing

You can even combine palm pressure from the cup, palm pressure from same hand fingers, and scrunching toes of both feet, Figure 11-6, while you are swallowing.

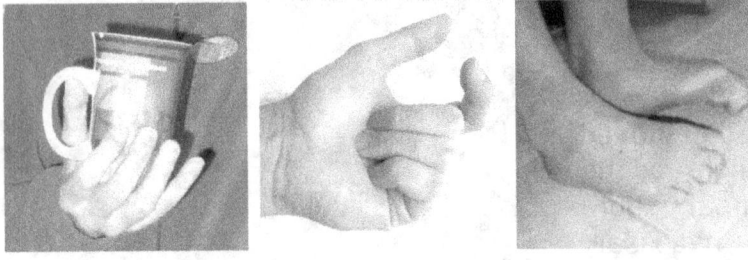

Figure 11- 6

Miscellaneous issues

Typing Techniques (that may help gain some finger control)

The fingers are often the first affected by Parkinson's and possibly the most affected. It is difficult to bring back enough dexterity and control to help enough using the techniques. But when my finger control is bad, 30 to 40% increase still does not bring my fingers to the dexterity level I used to have. Still, some techniques may help a little at times. It does help a bit though. Sensory Neural Pathway techniques to try: smiling, chewing gum or pretend to chew gum, scrunching toes, music with a beat, a head mounted laser pointed at the wall if your laptop is set up with the wall behind it, Figure 11-7.

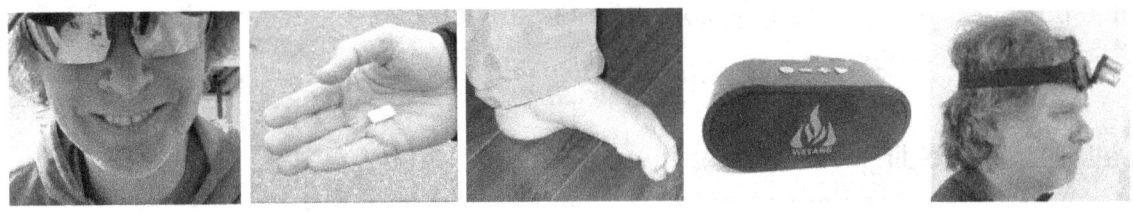

Figure 11- 7

When taking meds, if no water, try this

To bring saliva to your mouth chew gum and your mouth fills a bit with saliva that can be used for swallowing meds.
Another thing you can try is actually think the word '**salivate**' in your mind. Your mouth will actually fill with some saliva that can be used.
Test both strategies being sure to have with a glass of water nearby.

Rolling a yoga mat

Stimulate your neural pathway to make rolling up your yoga mat easier, tighter, and faster. Your hands should grip the yoga mat tightly each time you grab the mat, Figure 11-8. As you roll, you must consciously grab and squeeze a bit so that the neural pathway is active with that hand. Then just continue rolling this way until complete.

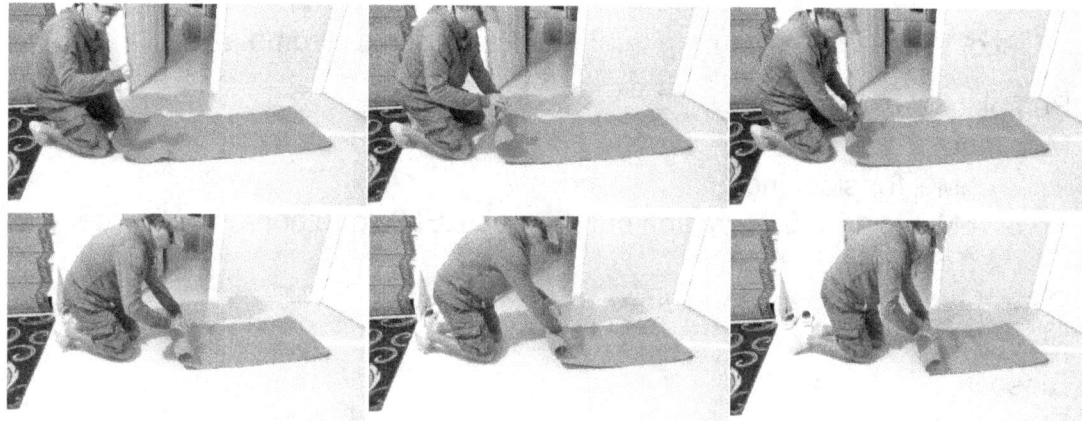

Figure 11- 8

<u>Tips to help Constipation</u>

With our muscles for digesting food receiving less messages from our brain as Parkinson's progresses, constipation becomes an increasing problem as waste moves more slowly through us. Practice my techniques often and I believe while the neural pathway is opened a little more it also must help the digestive muscles function a bit. This is my theory.

To help with constipation:

- Hard exercise daily if possible such as jogging, cross country skiing, boxing, rowing machine, exercise bike
- Sit-ups, daily if possible
- Prune juice
- Bananas
- Buy frozen prunes, boil, and store in a jar in the fridge
- Snack: sliced carrots mixed with grapes or sliced mango
- Extra Strength Dieters' Nature Slim Tea. Effective in helping constipation. Check Chinese store tea section.

For how to improve bowel movements using sensory techniques see *Chapter 12 - Personal Care, Bathroom, and Showers.*

Insomnia

Insomnia gradually became a nightly visitor. I exercised every morning on my treadmill out of fitness habit but this would also serve the dual purpose for attempting to get an early start at trying to tire myself out by end of day. To get to work I took the bus. I walked 2.5 km home for the same reason as using the treadmill. As insomnia grew exercise no longer was enough.

Here is a list of things that might help deal with Parkinson's insomnia:

- Pumpkin seed tea
- Melatonin
    - 5MG for sleeping
    - 5MG to 15 MG for acting out dreams/REM sleep behavior disorder (RBD)
- Over-the-counter sleep medicine
- Exercise daily
- Sleep in a dark room
- Eye-cover
- White noise

Vitamins

Before I was diagnosed with Parkinson's my doctor told me my Vitamin D was low. After learning I was low, I began taking Vitamin D and eventually I switched to Vitamin D3. I know now that low Vitamin D is common with people with Parkinson's. As Parkinson's advanced I looked into vitamins that were thought to help slow progression. When Dyskinesia and Dystonia started becoming a problem I researched even more, hoping to fill my brain with vitamins that would help slow progression or encourage dopamine production. I read books, searched on the internet for articles, read forums where others with Parkinson's wrote of their beliefs and experiences with vitamins. The truth is that I don't know if the vitamins are having an impact but I do know that I've progressed slowly.

Vitamins you might try:

- B3
- B12
- D3
- Turmeric/Curcumin
- Gingko Biloba

The reasons that I take these vitamins can be found by searching "Parkinson's" and the vitamin on the internet.

Chapter 12

Personal Care, Bathroom, and Shower

The shower

Showering can be difficult when Parkinson's symptoms are strong and arm/hand control is weak, Figure 12-1. Applying soap and scrubbing is made into a struggle while you try to move your arm and apply pressure to make your hand scrub. There are a few strategies I like to use.

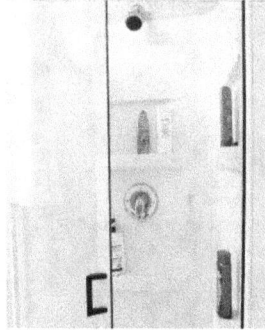

Figure 12- 1

Washing in the shower

When showering, many things can become a struggle if Parkinson's meds are not at full strength or not helping enough and the amount of hand/arm control you have is not ideal. Difficulty with initiation of motion, tremors, and slow movement might affect the washing process. Whether you are applying body wash, Figure 12-2, using a bar a soap, Figure 12-3, scrubbing with a washcloth, Figure 12-4, or applying shampoo/conditioner, Figure 12-5, sometimes it is difficult to move your hand(s) properly and to apply enough pressure.

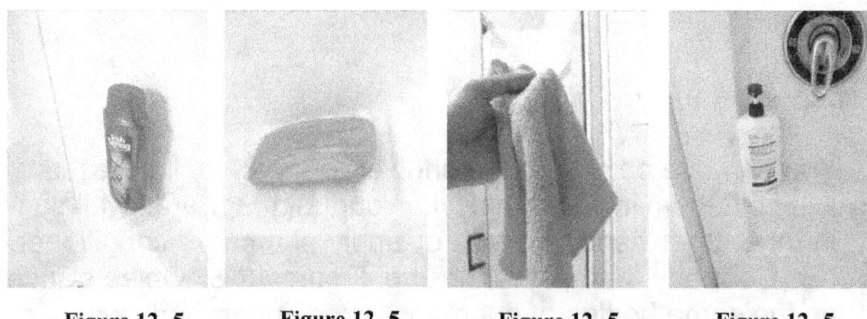

Figure 12- 5   Figure 12- 5   Figure 12- 5   Figure 12- 5

To help improve the process apply palm pressure to stimulate better movement and control in your other arm. I recommend applying the soap, shampoo, etc., and using the washcloth with the arm that you have the best control of and using the weaker hand to apply palm pressure. You will be using one hand for washing. The other hand will use palm pressure through the sense of touch and use of the hand's muscles to temporarily widen the neural pathways for messages to your hand that is doing the washing.

Techniques to improve hand control while in the shower:

Shower wall palm pressure

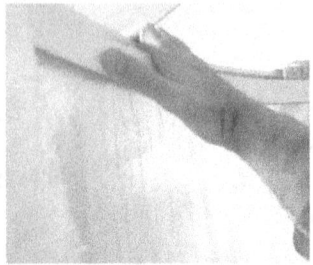

Figure 12- 6

While applying soap/shampoo/conditioner, using a washcloth or towel, if you are struggling to control your hands then put one hand up on the shower wall's top edge, Figure 12-6, and press downwards in a way that creates pressure on your palm. As an alternative reach upwards and press your up-wards reaching hand against the flat wall above you and press so that your palm feels pressure. Your other hand should feel increased control. Another benefit is that you will also only be focusing on using that one hand for washing movement.

Hand pushing on shower shelf

Figure 12- 7

If your shower has shelves make use of them when you need extra control of your arms/hands. Use one hand (I often use my weaker hand) to apply upwards or downwards pressure on the palm, if possible, or on the fingers, Figure 12-7. This will allow the other hand to move better and let you have more control for washing. Maintain the upwards pushing pressure while your washing hand is busy washing. Experiment with and without creating palm pressure to see the difference.

Hand holding a bar of soap in palm

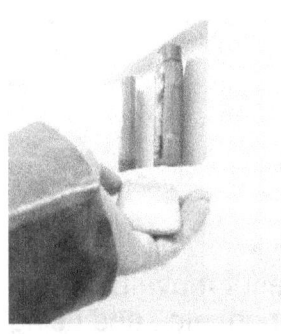

Figure 12- 8

You can use objects in the shower to create palm pressure, just hold them in your palm with your palm upwards while us-ing the other hand to scrub or apply shampoo, conditioner, soap, or body wash, or for rinsing, Figure 12-8. Objects such as the same bottles of shampoo, conditioner, body wash, or the hand soap can be held in the hand that is not doing the washing. A smaller bottle would be my choice if I use a bottle for palm pressure. Again, you should feel more control in your non-palm-pressured hand.

Note that any of these palm pressure techniques may help you move your legs if you happen to **freeze** in the shower.

Other Sensory Neural Pathway techniques to use while showering: smiling, chewing gum or pretend to chew gum, scrunching toes, music with a beat, Figure 12-9.

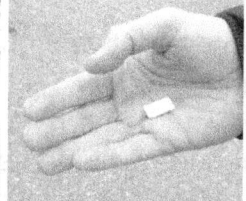

Figure 12- 9

Technique for combing hair or brushing teeth

When you brush your hair, hold the brush with your better hand and use it for hair brushing while simultaneously using a variation of the palm technique by placing your other hand under the counter and gently pushing upwards, Figure 12-10. It works the same for brushing teeth.

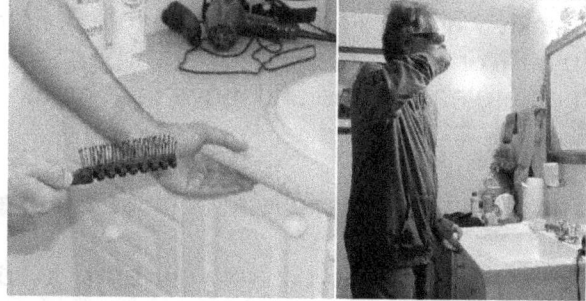

Figure 12- 10

You can also scrunch your toes on one foot or both feet.

Technique for blow drying hair

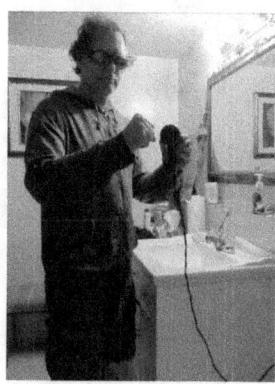

Blow drying your hair is trickier because both hands are occupied, one holding the blow dryer and the other holding a brush. The solution is to used palm pressure and sensory hand muscle movement with the hand that holds the blow dryer. Grip the blow dryer's handle tightly to do this, squeezing like making a fist, Figure 12-11. The other hand can hold the brush as you would regularly hold it, keeping it more flexible for styling, Figure 12-12.

Figure 12- 11

Figure 12- 12

Techniques for towel drying

Towel drying counter palm pressure

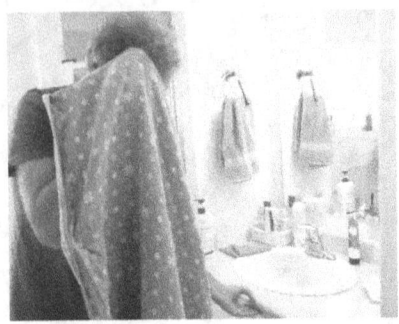

Figure 12- 13

Sometimes if you shower at your off med time, you will have difficulty drying your hair. A strategy to use is to hold the towel with your better hand and use it for drying while simultaneously using a variation of the palm pressure technique by placing your other hand under the counter and gently pushing upwards while you dry yourself with the towel, Figure 12-13.

You can also scrunch your toes on both feet. Another option is if you have gum in your mouth is to chew it quickly while you are attempting to use both hands to dry your hair. You will gain up to 30% or 40% more control over your hands while you chew. If you do not have gum then pretend you do and move your mouth in a chewing motion.

Towel drying neck movement muscle movement

When struggling to control your arms for drying your back, try this trick to increase arm mobility. As you pull the towel to your left, simultaneously turn your head to the left, Figure-12-14. Then pull the towel to the right while turning your head to the right. Repeated until you are dry. Moving the neck muscles has the impact of widening the neural pathway for better messaging to the arms, resulting in increased arm control.

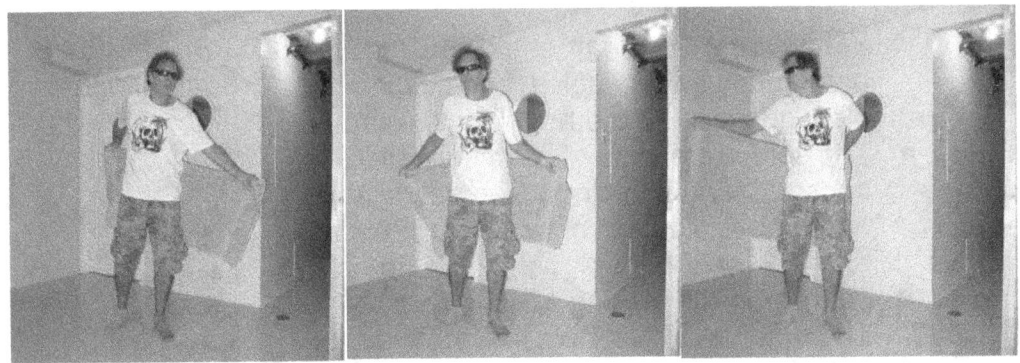

Figure 12- 14

Towel drying fist grip

Hold the towel tightly with one or both hands in a tight fist-like grip, Figure 12-15 while drying yourself. This works for drying your back, hair, front, arms, legs, etc. It is a little awkward but it does increase mobility, allowing you to better control arm movement and to use increased pressure for drying.

Figure 12- 15

Other Sensory Neural Pathway techniques to use while towel drying yourself: smiling, chewing gum or pretending to chew gum, scrunching toes, music with a strong beat, placing the laser pointer down on a counter so that the laser points on the wall where you can easily see it while you are drying, Figure 12-16.

 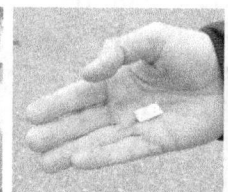 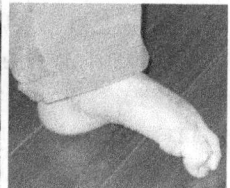  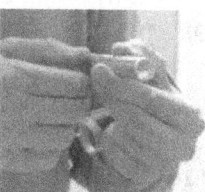

Figure 12- 16

On the bathroom throne - improving bowel movements

Parkinson's weakens signals to the muscles throughout our body.

This would include the muscles involved when we sit on the throne and push to empty our bowels, Figure 12-17. It is more difficult to know if palm pressure helps for different reasons. When palm pressure is used while we walk our gait visibly changes. There is no way to visibly measure effects of palm pressure while on the throne. Passing stool is also related to the digestive system and how well our system processed and moved the food over time. That said, I do believe palm pressure does impact the messages to the muscles all over our body including those involved in pushing stool out.

Figure 12- 17

Toilet palm pressure technique

Using something as simple as a bread bag plastic clip to hold in your palm while sitting on the toilet should open the neural pathway and help increase the push, Figure 12-18. While sitting there doing your business try to be aware of the feeling of force from the muscles involved in moving stool. Without changing the amount of force, pick up the bread bag plastic clip and place it into the palm of your other hand. That arm should be bent and forearm parallel to the ground with palm facing upwards. Notice that the feeling of force in those muscles moving stool suddenly feels a little stronger. This is the effect of the neural pathway opening a little wider.

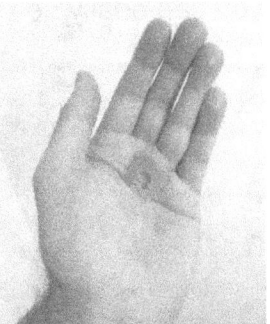

Figure 12- 18

Original palm pressure technique

The original palm pressure technique can also be relied on, Figure 12-19.

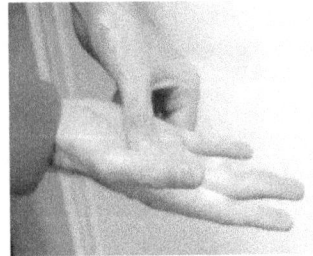

Press the thumb of one hand into the palm of your other hand and note the sensation of increased pushing at your bowel muscles. It is a noticeable improvement but remember it does not go back to normal non-PD level. The improvement is welcome though.

Figure 12- 19

Toilet - palm pressure with chin on palm

You can create palm pressure in a convenient way when sitting on the toilet. Rest
your elbow on your knee and rest your chin
on your palm, Figure 12-20. This creates
pressure on palm or palms. You should notice a difference. Push without palm
pressure and without altering the strength
of the push, do the elbow-on-knee-and-
chin-on palm technique. You should notice
that it feels like the push muscles involved
are more effective.

Figure 12- 20

Toilet - lift your foot

While sitting on the throne and taking
care of business, begin with your foot
flat on the ground. Next raise the ball
of your for foot off the ground, Figure
12-20. Either hold or repeat. You
should experience an improvement in
your bowel muscles' performance.

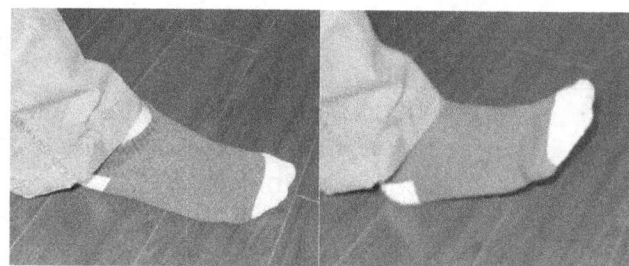

Figure 12- 20

Toilet - kick your leg out and back

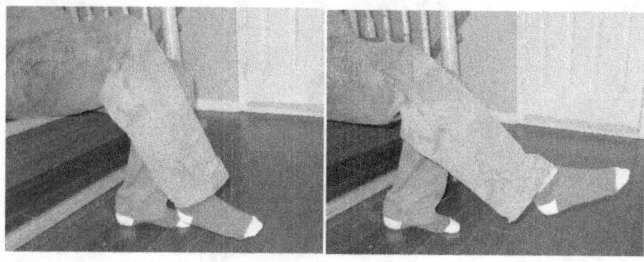

Figure 12- 21

While on the throne, kick your leg
out, hold position for a second or two
and swing it back, Figure 12-21.
You should feel an increase in pushing pressure magically appear.

Toilet - other techniques:

Other techniques you can try in conjunction with the toilet techniques or on their own while you sit on the bathroom throne: smiling, chewing gum, scrunching toes, music with a beat, setting a laser pointer down and having it point to a visible location, Figure 12-22.

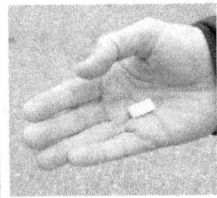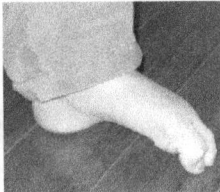

Figure 12- 22

Ultimate combination for the toilet
- Chin on hand palm pressure
- With laser pointing at floor or wall
- Strong beat music playing

Chapter 13

How to use Palm Pressure in Everyday Life

The palm pressure technique, Figure 13-1, is where you press your thumb into the palm of your other hand, creating a message from your brain to your palm that seems to open the nerve message pathways a little more for messages from your brain to your legs, resulting in a smoother heel-to-toe stride.

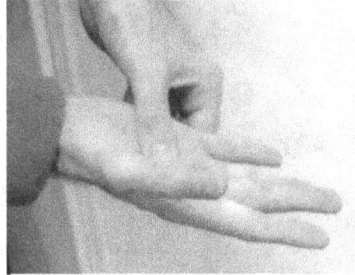

Figure 13- 1

When you think about it pressure on the palm of the hand can come from many objects throughout the day. Your hands are not always both free to use this technique where you put the thumb of one hand into the palm of the other hand but the very reason that they are not both free may present an opportunity to use a variation of it. You just need to be aware of the opportunities that present themselves and build the habit of carrying objects differently and of making use of objects.

Palm pressure techniques with everyday objects

The following are examples of things you might carry using this palm technique. You should notice the change in your stride while you carry objects in your palm. When the change in stride control happens use it to lengthen your stride, rather than to try to take quicker steps.

Carry a bottle cap

Just to demonstrate how sensitive the palm of your hand is, place a bottle cap into the centre of your palm, Figure 13-2, and walk. Remove the bottle cap while continuing to walk and notice that you stride changes, becomes slower, and may shuffle with toe-to-heel steps. Without stopping, put the bottle cap back into your palm and notice the improvement.

The bottle cap weighs practically nothing but when you put it into your palm, the brain automatically sends a message to your hand that it must stay busy holding the cap. It is a consistent message through the neural pathway that I believe not only opens the pathway more, allowing clearer messages to the legs to get through to muscles all over the body but with this constant message by holding up the bottle cap, it helps keep the passageway open for the duration.

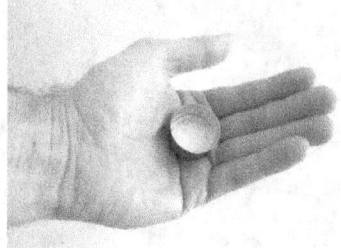

Figure 13- 2

Carrying a bottle cap also demonstrates that it is not the amount of pressure on the palm that is important but simply that your hand is aware something is on the palm.

Carrying a plate of food using palm pressure

You could be carrying a plate of cool carrots. Out of habit many carry a plate by pinching the edge as in Figure 13-3.

Figure 13- 3

Figure 13- 4

If we PD people change our habit to holding the plate in the palm of our hand, Figure 13-4, we will walk better and more heel-to-toe. This is an opportunity use the technique. Hot plates, of course means you'll need an oven mitt or maybe you must carry it the regular way. Still, think about it when you get a bowl of cereal or ice cream, etc. You could also carry the hot plate with palms upward, using both hands where the hands are just holding the cooler outer edge.

Any time we can take advantage of objects that we carry using the palm technique we can walk a little better and that just feels nicer.

Carrying a glass of water

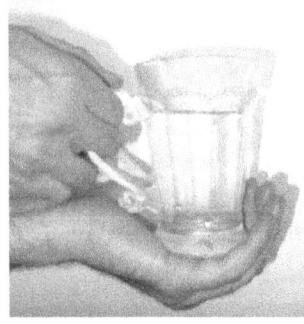

Figure 13- 5

Maybe you just went to the kitchen for a glass of water, so take advantage of having the cup and while holding it with one hand, carry it in the palm of your other hand, Figure 13-5. Your gait will improve during this time. Of course be careful if it is hot. Just make this minor adjustment when carrying a glass of wine, a bottle of beer, or a glass of prune juice and no one else will notice. Build the habit to make it automatic. Instead of struggling to walk with the cup or glass you might look forward to it.

Carrying a plastic bag

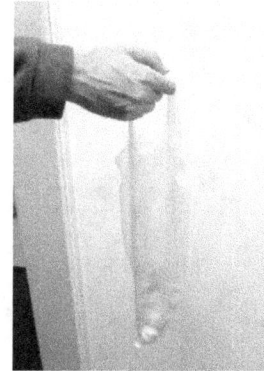

Even carrying a plastic bag can provide an opportunity to use the palm technique. Instead of carrying the bag in the traditional manner, Figure 13-6, we need to adjust and palm the bag, as in Figure 13-7.

It is a little more work but for short distances you should be okay. Walking smoother is the benefit and you will feel more normal.

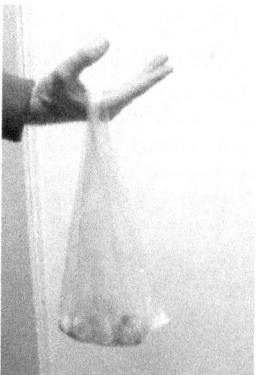

Figure 13- 6

Figure 13- 7

Bottle palm pressure

Maybe you like wine and need to go get a bottle from another room. Use the palm pressure technique to walk to get the bottle of wine and return carrying it with the bottle bottom in your palm, Figure 13-8.

Change the way you carry bottles of relish, mustard, ketchup, and jars of peanut butter. Every moment you have little victories and feel more normal and usual is good for moral.

Figure 13- 8

Chip bag palm pressure

Something as simple as carrying a bag of chips so that your palm feels the chip bag as well as feels the weight of the chips, small though it is, triggers the neural pathway to allow walking messages to get through more, Figure 13-9. This should improve your gait by 30 to 40%. At the very least you should experience noticeable positive change during the time that you carry the bag of chips in the palm pressure way.

Figure 13- 9

Laundry basket palm pressure

Adapt the way you carry a laundry basket, Figure 13-10. Use an open palm on the side grips of the basket as you walk. Every little bit helps to make you feel a little bit more normal.

Figure 13- 10

These are just some examples of every day occasions where you can incorporate the palm technique. If you go to get a cup of tea, carry the cup of hot water by the handle and carry the tea bag in the palm of your other hand until you reach your intended place to sit down. You can experiment and find new uses that help make your Parkinson's life a little easier. You will feel like you are winning little battles with PD even as it advances.

Long sleeve pressure

If you are wearing a shirt, sweater, or jacket with a slightly long sleeve that reaches the bottom of your palm, Figure 13-11, you can use the sleeve to help trigger improvement in your stride as well as a better arm swing. Allow the sleeve to reach your palm and then close your hand on it. This is enough to activate palm pressure, to help your gait improvement, and help improve your arm swing.

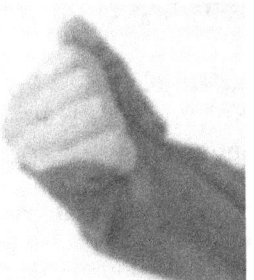

Figure 13- 11

Neural Pathway Enhancers

I have experimented with ways to stimulate the neural pathway. You can't always carry around a bowl, cup, wine bottle, hiking stick, etc. And for some exercise, even the walking pole is not convenient. So I adapted and made something easily portable that can be used with similar results. Using a mop or broom handle, Figure 13-12, that you can spare, cut pieces of varying lengths to make what I call Neural Pathway Enhancers (NPEs), Figure 13-13.

Figure 13- 12

Figure 13- 12

The shortest one I use indoors, often just carrying it around in my jacket or sweater pocket, ready for use if PD gives me trouble. I also use it for exercising and balance practice, even for punching on the small soft bag that I introduce in Chapter 16. In warmer weather I'll also use the two small wooden NPEs for walking. I will also combine using one of them in one hand while holding a hiking pole in the other hand. Both the hiking pole and the NPE trigger better messages to the legs and arms and you should walk smoother and have a bit more control of your arms and arm swing.

The medium length NPE I use in the winter. It is longer so that I can store it in my winter coat pocket and still be able to reach for it easily enough with my winter mitts because it sticks out of my pocket enough. In one hand I carry the NPE, in the other hand I carry the hiking pole.

Figure 13-13 shows examples of using the small NPE with a hiking pole and Figure 13-14 shows an example of using the medium NPE with a hiking pole.

Figure 13- 13

Figure 13- 14

Some days my arms are not cooperating as well as I would like and I'll use the small NPE to improve my strikes while doing a Sensory Overload MMA or Bike and Box workout. I just hold an NPE in each hand, making a fist around them, squeezing a little and I have better control over my arms (and legs), Figure 13-15. I get a more satisfying workout this way.

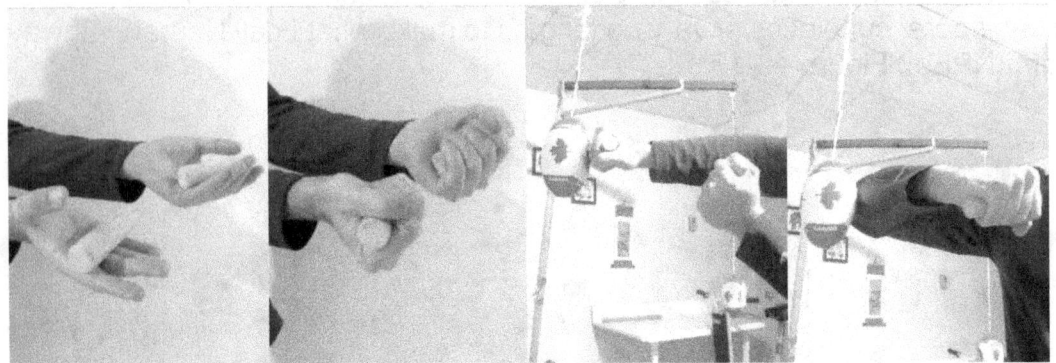

Figure 13- 15

The longest NPE that I use is about shoulder width length. It too is made from a wooden mop or broom handle. I either hold it palm facing downwards or palm facing upwards at stomach level, Figure 13-16. To maximize effect, I squeeze my grip a little and apply a bit of force as though trying to bend the NPE. It can be used in some exercises, such as the balance exercise where you step one foot in front of the other, raising your knee up to the NPE, before placing that foot directly in front of the other, Figure 13-17.

Figure 13- 16

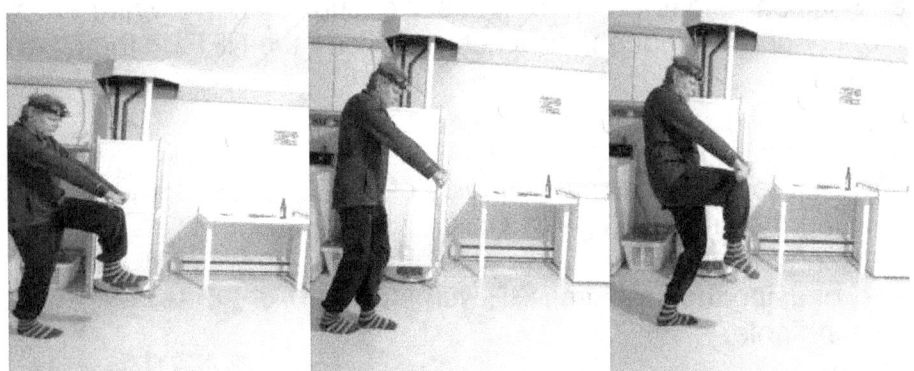

Figure 13- 17

Chapter 14

Balance and Balance Exercises

Once the stage of Parkinson's arrives where slow movement becomes a frequent issue, you may not realize it but your balance has also been decreasing. This will gradually impact standing and walking. You want to keep walking for fitness, mood, and digestion reasons. Walking helps keep the feeling of independence. In this chapter, I'll cover techniques that allow you to have better balance, so you are more stable in whatever you are doing, while using the techniques.

Balance practice and exercises

Better Balance, Better Equilibrium by Finger-snap

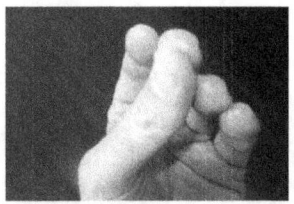

Figure 14- 1

Snap your fingers, continuously to trigger better muscle control and therefore better balance for the duration of this exercise.

First, for comparison, without using the technique, while standing, Figure 14-2, lift one leg up and try to maintain your balance. Next begin again only this time use the finger-snaps. From a standing position, begin snapping fingers, Figure 14-3. While continuing to snap lift the same leg. You should notice an improvement in balance. Attempt with and without the finger-snaps several times to confirm the difference. Maintain balance longer is good training for the muscles involved. You also see how the techniques help with balance issues and how they can benefit your gait. Your steps will shorten if you are losing balance while walking. Increased balance allows for a better, longer, more in control stride.

Figure 14- 2   Figure 14- 3

Balance front-kick exercise - thumb-press

From a standing position with right leg behind your left leg, slowly kick your right leg forward as though you are doing a kick to the groin, Figure 14-4. Once extended, hold for a few seconds, maintaining your balance. It is not necessary to kick completely at groin height, the objective is being able to hold the position. Repeat, kicking with the left leg. For increased balance use the thumb-press technique to stimulate the neural pathway.

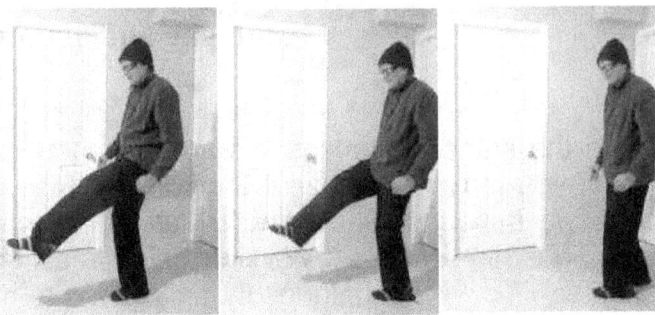

Figure 14- 4

Balance exercise - calf raise and step

For this balance exercise, first attempt it without any techniques. Stand with one foot directly in front of the other foot. Lift the back foot, raising it the foot so that the heel of the foot rises towards your buttocks. Raise it about halfway. Then move that leg, stepping forward and placing that foot directly in front of the other foot. Now repeat with the other leg. Keep stepping forward in a straight line, maintaining balance as best you can, Figure 14-5.

Figure 14- 5

For better balance try holding a sawed off broom handle while you do the balance exercise, Figure 14-6. Squeeze it tightly and even apply some force as though you are trying to bend it. Your body control and balance should improve for the duration.

Figure 14- 6

Balance step exercise - palm pressure using dumbbells

First hold the dumbbells lightly in your hands, palms facing downwards. Walk, stepping one foot in front of the other, Figure 14-7. Do not squeeze the dumbbells. Be aware of your level of balance.

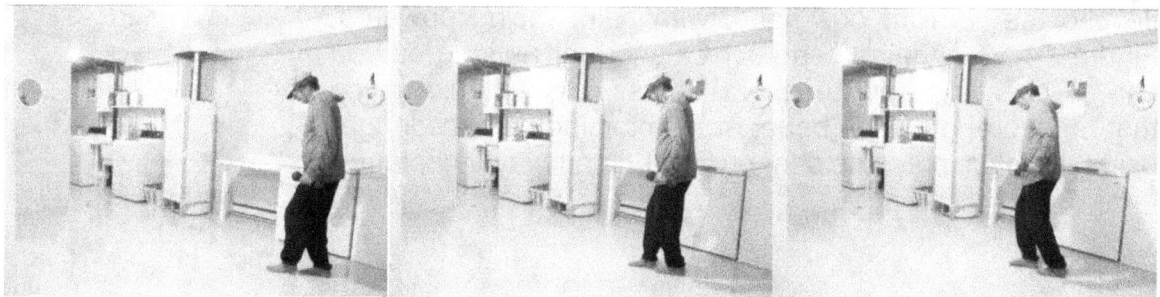

Figure 14- 7

Now repeat, this time holding the dumbbells palm upwards and even squeezing them a bit, Figure 14-8. You should feel more control and better balance. You may still lose balance occasionally but it should be less often than when not using this technique (or other techniques).

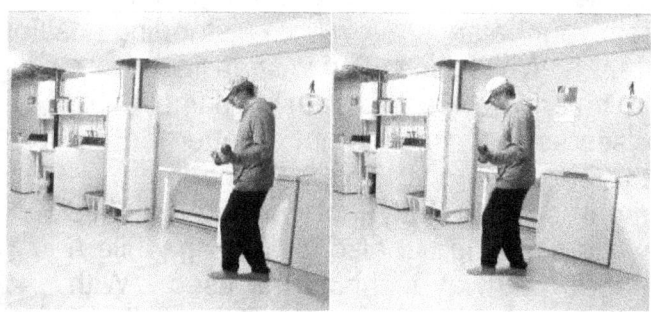

Figure 14- 8

Balance one-leg-standing exercise - palm pressure using dumbbells

First, attempt this exercise with noting in your hands. From a standing position, with hands raised palm upwards and nothing in them, lift one leg and bring the heel of that leg up to the knee of the other leg. Maintain this position as long as you can, testing your balance. Now repeat only hold a dumbbell palm upwards in each hand and maintain balance as long as you can, Figure 14-9. Over the course of five or more attempts with and without dumbbells you should notice that palm pressure by dumbbells leads to better balance. You can also try using soda cans or quarters, for example, instead of rocks. This is a good indicator of how palm pressure helps with balance. It also helps with increase balance and speed when walking.

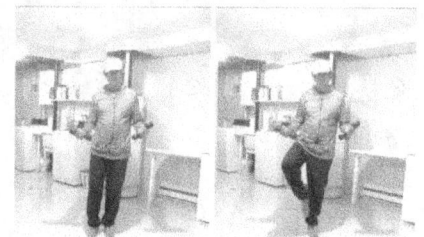

Figure 14- 9

Balance lift-one-leg exercise - palm pressure using a med bottle

For this balance exercise, first without using any sensory tricks, while standing, lift one leg a few inches off the ground and maintain your balance for as long as possible. Then try again, only this time place a medication bottle or vitamin bottle into the palm of one hand, with palm facing upwards, Figure 14-10. Again keep your balance for as long as you can. Repeat both ways several times. You should notice that on average you do better at maintaining balance longer using the palm pressure sensory neural pathway technique.

Figure 14- 10

Balance stand-raise-one-leg exercise - finger-snap

While standing, without using any sensory techniques, lift one leg up and try to maintain your balance. Now from a standing position, begin snapping your fingers, Figure 14-11. While continuing to snap lift the same leg. You should notice an improvement in balance. Repeat both standing exercises several times to get a better feel for the improvement. This balance improvement should apply not only to balance while standing but also to walking while a neural pathway technique is used. With better control of balance while walking, your gait should benefit by becoming a little smoother. As always the benefit is based on your current balance and gait that Parkinson's and Parkinson's medication are allowing you at any point in time.

Figure 14- 11

Balance short-step exercise - thumb-press

To improve balance, press your thumb up against your index finger continuously as though you might flick your thumb but don't flick, instead maintain pressure. Step forward, one foot placed directly in front of the other, going in a straight line, Figure 14-12. Then repeat without using any sensory technique. Compare with and without the technique. Again you should notice an improvement in balance when using the thumb-press technique.

Figure 14- 12

Balance couch-to-stand - thumb-press

To demonstrate how the neural pathway techniques help with balance I go from sitting in a low couch to standing several times without using and then using the thumb-press technique. I use the same crossing of my arms standard balance test the only difference is in the 'After' I press my thumbs against my index fingers. Both my balance and speed of movement improve by 30 to 40%. This also hints at how neural pathway tricks can help balance in general for activities like walking.

First, without using a neural pathway trick, while seated in the couch, cross your arms, reaching towards the opposite shoulder, Figure 14-13. This way arms will not be used in a balancing way or to push off from the couch. Now stand. You might struggle a bit.

Figure 14- 13

To test the effect of tricking the neural pathway, use thumb-press technique. This technique is done by pressing the thumb into the index finger continuously, Figure 14-14.

Figure 14- 14

Now repeat this balance test but this time press your thumbs against your index fingers, Figure 14-15, the entire time that your arms are crossed. You should find that you stand up more easily and quickly.

Figure 14- 15

Balance lift-one-leg exercise – Finger-snap

For this balance exercise, without using any sensory technique, while standing, lift one leg a few inches off the ground and keep your balance for as long as you can. Repeat only this time begin snapping the fingers of one hand or both hands before you lift up one leg and continue for the duration of time that you manage to maintain balance, Figure 14-16. You should notice over the course of repeating that you have better balance when using the finger-snap technique.

Figure 14- 16

Balance short-step exercise - headlamp

Using a headlamp, Figure 14-17, or holding a flashlight, aim the light at the floor several feet in front of you. Step forward, placing one foot directly in front of the other doing a straight line, Figure 14-18. Remember to be aware of the light on the ground. You do not need to stare directly at it. Try the same balance exercise without the light. You should notice that you maintain better balance and possibly have more speed while using the light.

Figure 14- 17

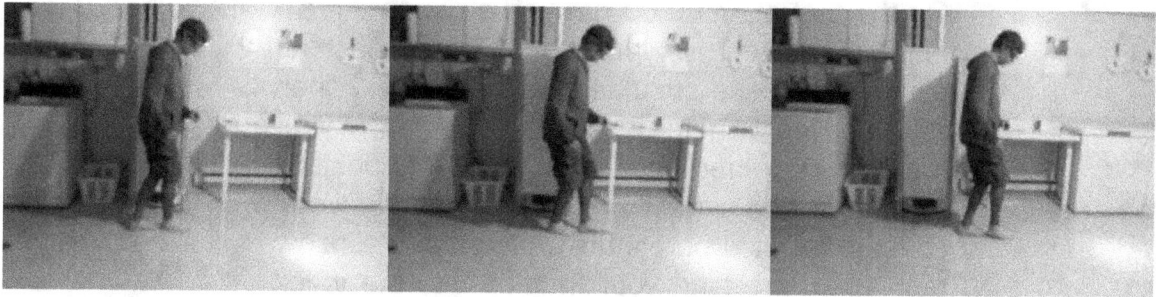

Figure 14- 18

You can attempt the same short-step balance exercise using a head-mounted laser, holding a laser pointer, or while listening to music with a strong, steady beat, Figure 14-19, and you should also see improvement.

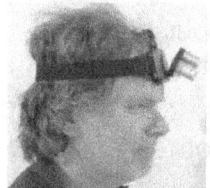

Figure 14- 19

Sawed piece of broom handle

Another method to try the short-step balance exercise is with a sawed off piece of a broom handle (or the entire broom if you have room). As you take steps hold the broom handle piece, hands space shoulder width apart and either palms upwards or downwards, Figure 14-20. Squeeze the handle and even apply a little pressure as though you are trying to bend the handle.

Figure 14- 20

Balance one-leg-standing exercise - thumb-press

For this exercise, press your thumb into your index finger continuously for the duration, Figure 14-21. Do this with both hands. I don't think using the neural pathway technique with both hands increases the benefit but I do think it maximizes the stimulation to the neural pathway in case using one hand does not quite manage to get maximum stimulation.

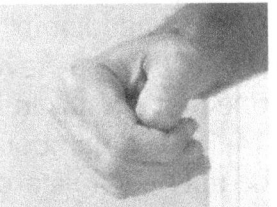

Figure 14- 21

Figure 14- 22

On the first attempt, do not use a sensory technique. Simply cross your arms, fingers reaching for your opposite shoulder, Figure 14-22. Then raise one leg, bending it and place the heel of your foot on your other leg's knee. Hold this position as long as you can. You may wish to try several times. Now incorporate the thump-press sensory technique, pressing your thumbs into the index fingers while keeping your arms crossed, Figure 14-23.

Figure 14- 23

Balance by sidekick

Practice balance and strengthen muscles involved in maintaining balance by doing martial arts sidekicks, Figure 14-24. From a standing position, lift your right leg so that your knee is nearly hip height. Next kick your right leg out to the side so that your leg is approximately hip height. Hold for a few seconds if you can. In the beginning as you practice maintaining balance and building muscle strength in both legs, including your hips you might only be able to kick out at knee height, don't worry. Repeat several times. Then do this exercise this time kicking to the side with the left leg.

Figure 14- 24

Balance by roundhouse kick

Another good balance exercise kick is the roundhouse kick. Lift you're your right knee while turning the supporting foot and body in a semicircular motion, Figure 14-25. Extend the right leg as though striking with the top of your foot or with your shin.

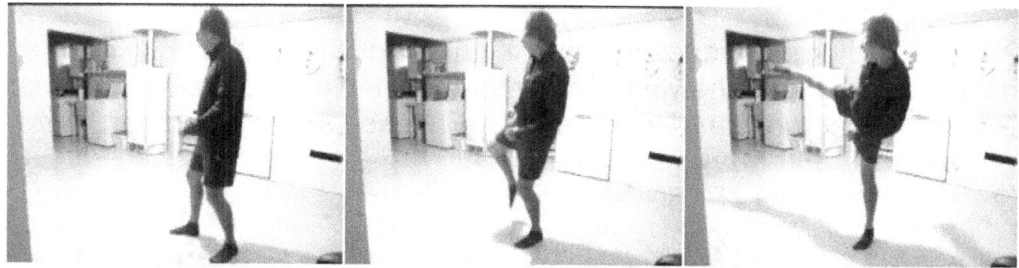

Figure 14- 25

Shoe target practice

Gather many shoes (or other objects) and space them on the floor so that each shoe-to-shoe distance represents a long stride, Figure 14-26. The objective is to walk, stepping so that your foot lands beside each shoe. Begin walking several feet away from the first shoe, with your regular walk. Once you reach the first shoe, aim the next step for shoe number 2, which, like all the shoes, has been spaced at a distance for a long stride. You will instinctively and easily step longer. Step along through the shoe course. Not only does your stride become longer, smoother, and more heel-to-toe but you might notice your arms relaxing a bit and moving a little more freely. This exercise will be most beneficial when your natural Parkinson's stride is shorter and slower, a shuffle.

Figure 14- 36

The shoes act as targets for your feet to try to reach. This exercise allows your body to practice more normal movements.

Other techniques to help with balance exercises:

Figure 14-28 shows other techniques you can try in conjunction with the techniques above or on their own: smiling, chewing gum, head-mounted laser or hand-held laser pointer, music with a beat, wooden Neural Pathway Enhancers (see *Chapter 13 - How to use Palm Pressure in Everyday Life*).

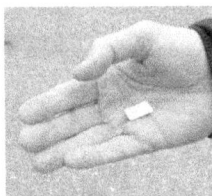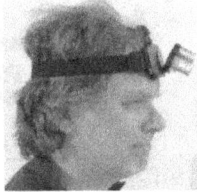

Figure 14- 28

Footprint training room

It would be great if Parkinson's groups could set up an empty room with footprints painted on or stuck on the floor, Figure 14-28. Because PWP do not all have the same length of legs or ability to walk, each set of foot prints would be of different stride lengths, with the emphasis being on longer strides. PWP can practice a longer stride matching their steps with the footprints. The footprints act as natural targets. Your feet almost instinctively try to place themselves on the footprints using this sensory trick. You should also notice a loosening in your arms and a better arm swing motion.

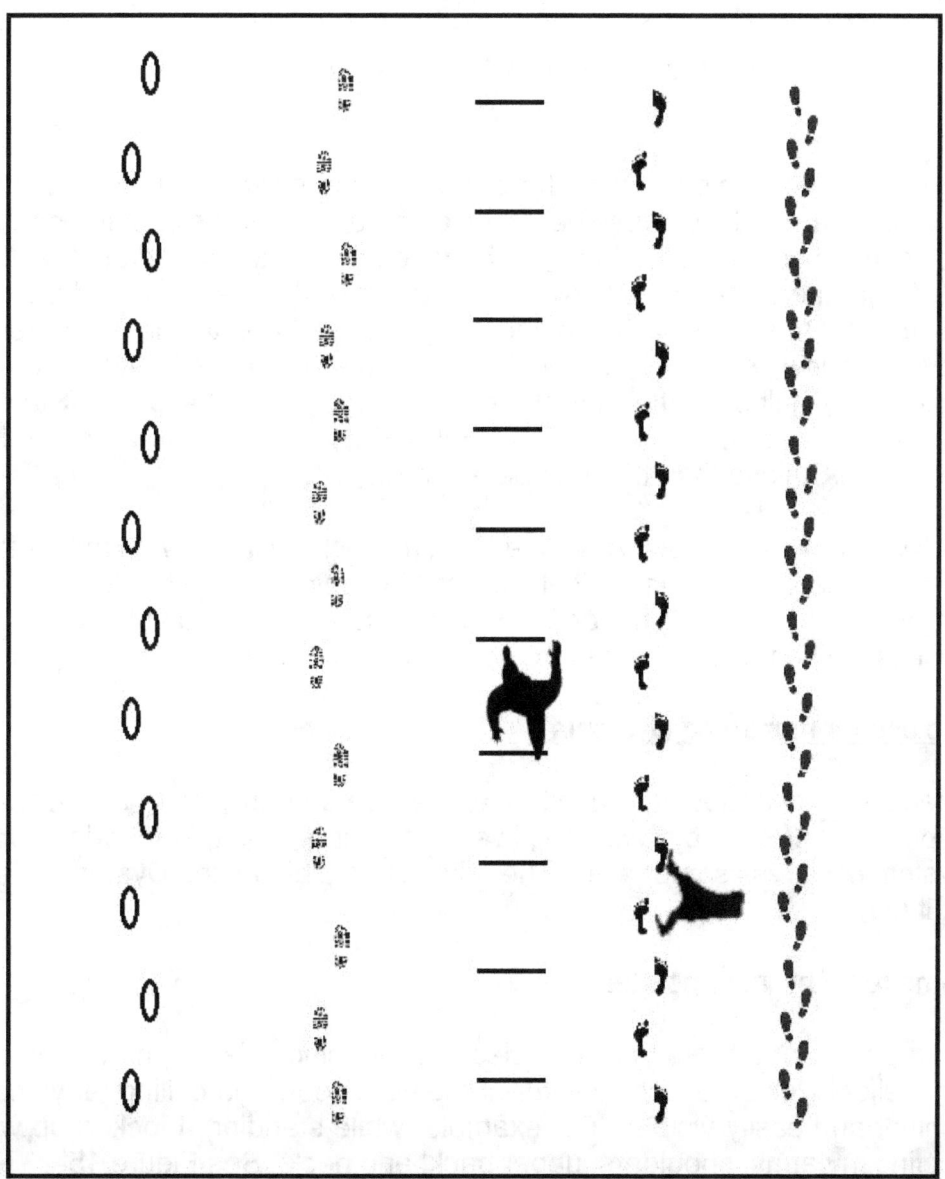

Figure 14- 28

Chapter 15

Dystonia and Dyskinesia

Once the stage of Parkinson's arrives where movements slow down life becomes difficult. Just walking around the house becomes a chore and chores become harder. Just standing up may become something you have to prepare to do. Add to that that many of us PWP get the side-effects known as Dystonia and Dyskinesia and life can become miserable at times. Likely this happens if you have been taking either Levadopa or Stalevo for a few years. As no med stops these side-effects from happening and Amantadine only helps a little, my strategy was to reduce Levadopa/Stalevo doses and some days not take any dose. Note that if you can switch to something other than Levadopa/Stalevo then that can be explored.

If you are stuck with Levadopa/Stalevo and reduce, my tricks will help with movement and mobility. The tricks can also help with Levadopa/Stalevo induced Dystonia giving some temporary relief by reducing the amount of Dystonia. You'll still suffer but it will be noticeably less. I use tricks to help manage Dystonia when it is at its strongest.

Techniques to use for managing Dystonia

Aside from medication changes and monitoring exercise so that you do not overuse muscles that become affected by Dystonia, I've found certain things that bring some relief from Dystonia. I stress some relief. The relief helps a bit but the Dystonia is still strong. Still... it helps.

Dystonia - some relief by laser pointer

When Dystonia strikes and locks your muscles up, try using a laser pointer, Figure 15-1, for some relief. Point the laser pointer at the wall, floor, and ceiling, anywhere that is convenient and easily visible. For example, while standing, I locked up with Dystonia affecting my arms, shoulders, upper back, and neck. See Figure 15-2. Notice how far my neck is pulled forward. I then shine the laser pointer light at the floor. Immediately my arms, shoulders, and back loosen. My head moves upwards as the neck muscles that pulled it downwards loosen too.

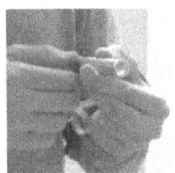

Figure 15- 1

Figure 15- 2

Dystonia – music with a beat provides some relief

Figure 15- 3

Music with a strong beat also can have the effect of temporarily reducing Dystonia, Figure 15-3. To demonstrate the strong beat in another way, I begin standing with Dystonia, particularly affecting my left arm, Figure 15-4. Focus on the stiffness and how my left arm hangs. With my right arm, I begin to slap the table in front of me, Figure 15-5. There is a loud smacking sound each time that I slap the table. Notice how my left arm drops bit by bit. In real time this happens relatively fast. Dystonia had been putting my left arm in the original tight muscled muscle position. The sound of smacks, like a beat, stimulated my auditory sense and my arm muscles relaxed. Possibly there may have been some benefit of right arm muscle movement and from palm-pressure upon slapping the table. You can do the same experiment playing music with a strong beat. Remember it helps relieve some Dystonia, it does not relieve all. Some is better than none when Dystonia hits.

Figure 15- 4

Figure 15- 5

Dystonia – palm-pressure provides some relief

Though using palm-pressure is less ideal than using a laser pointer or listening to music as you try to relax your Dystonia'd body, Figure 15-6 demonstrates that other sensory techniques may help reduce the suffering that Dystonia causes. Notice that once the soda can is placed in the palm of my hand, my left arm automatically drops down a bit.

Figure 15- 6

Chapter 16

Exercise for Parkinson's

Once the stage of Parkinson's arrives where our bodies slow down more often and things like getting off the couch or bed, or turning in bed become difficult daily events, we can benefit by keeping our arms strong, targeting the muscles involved in helping with our difficulties.

Stretching and Strengthening

Triceps curls

With five pound dumbbells and laying on the floor, we can do many repetitions of triceps curls, targeting the muscles which are muscles used when pushing off the couch and often also used in maneuvering in bed, Figure 16-1. From a laying down on your back position, begin by holding the dumbbells straight up, palms facing the same direction your head is. Bend at the elbows and lower the dumbbells until your forearms arm parallel to the flower. Return to the start position with your arms straight up. Do many repetitions. This exercise helps keep our triceps stronger. We use our triceps for maneuvering and shifting position in bed.

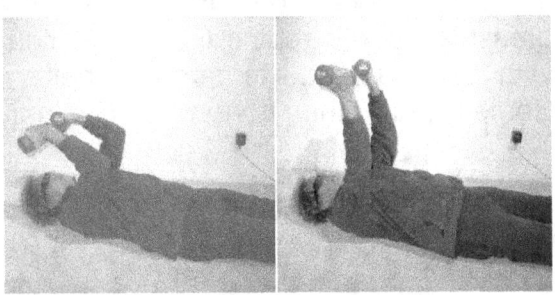

Figure 16- 1

Bench press

Another good exercise to help in the push off/up from a bed or a seated position is bench pressing five or ten pound dumbbells, Figure 16-2.

From a laying down position hold the dumbbells straight up. Next lower your arms so that your elbows are near or even touching the floor. Push upwards until your arms are straight again. Do many repetitions.

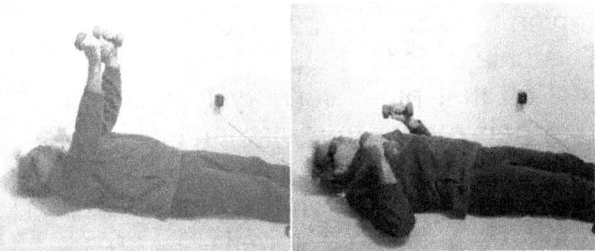

Figure 16- 2

Head and leg raise

To keep the back of the neck strong and work on keeping a more proper neck/head posture, I like doing simultaneous head and leg raises.

From a hands and knees position on the floor, raise both your head and left leg, Figure 16-3. Then lower both again. Repeat 20 or 30 times with the same leg. Then do the same with your right leg. This exercise is also good for the glutes which don't always get enough strengthening attention.

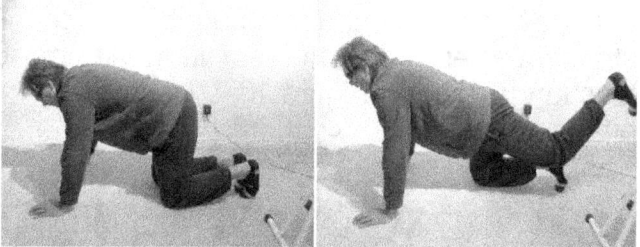

Figure 16- 3

Marching in place

Figure 16- 4

To keep legs and arms loose and 'remembering' they can extend far, practice marching in place with exaggerated movements. Lift your left leg high while swinging your right arm upwards and high too, Figure 16-4. Then do the same with your right leg and left arm. Continue marching in place. It helps keep your joints loose as well as helping keep muscles in the arms and legs strong.

BIG steps

Walk a straight line while taking exaggeratedly long steps, Figure 16-5. Similar to marching in place only you are advancing by taking long strides. Exaggerate your arm swinging motion to both the front and back. The idea is to keep flexibility and muscle memory.

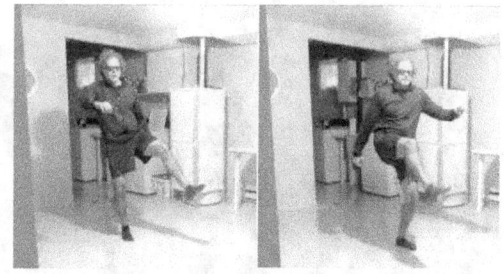

Figure 16- 5

BIG steps with ankle weights

To build strength in muscles involved in stepping forward with longer strides, try doing the same BIG steps while wearing ankle weights Figure 16-6. I wear 5 pound ankle weights but two or three pound weights would offer good resistance. The idea is to keep stride muscles strong and remind our leg muscles that we should take longer strides, Figure 16-7.

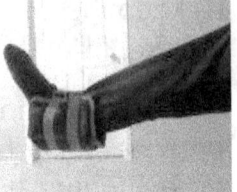

Figure 16- 6

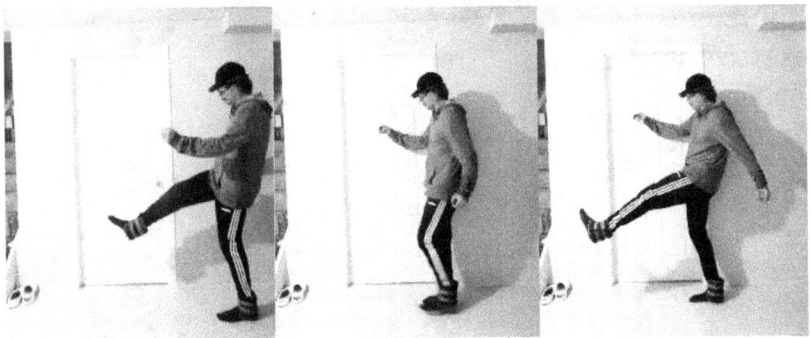

Figure 16- 7

BIG crossovers

To do BIG crossovers, begin with legs apart at shoulder width and arms spread very wide, Figure 16-8. Cross your left leg in front of your right leg while you simultaneously cross your left arm in front of your right arm directly in front of your body. Continue the sideways motion by moving your right leg to the right, effectively uncrossing your legs while once again spreading your arms wide. Continue in this direction only this time cross your left leg behind your right leg and your left arm behind your right arm. Repeat alternating crosses as space permits. Then reverse direction going towards the right. This is exercise good for arm flexibility, coordination and balance.

Figure 16- 8

Sensory Overload MMA

Sensory Overload MMA is an exercise that I created using two mini punching bags, Figure 16-9. I actually hang one from a ceiling vent at head height and the other at stomach height from my Bike and Box stand. You can choose the heights that you like. I do some kicks too and many I like to do on the lower bag but you may prefer heights of your choosing, maybe both set at head height. I alternate striking the bags as best I can, occasionally blocking as though the swinging bag is someone trying to strike me. I also duck and sidestep. The purpose is to let yourself get lost in the action, and trying to let instinct take over. You will be practicing initiation of movement through reaction and the goal is not perfection in striking and blocking but to just feel free, loosening your limbs up in a reactive way. It is fun and I sometimes imagine that I'm in a martial arts movie.

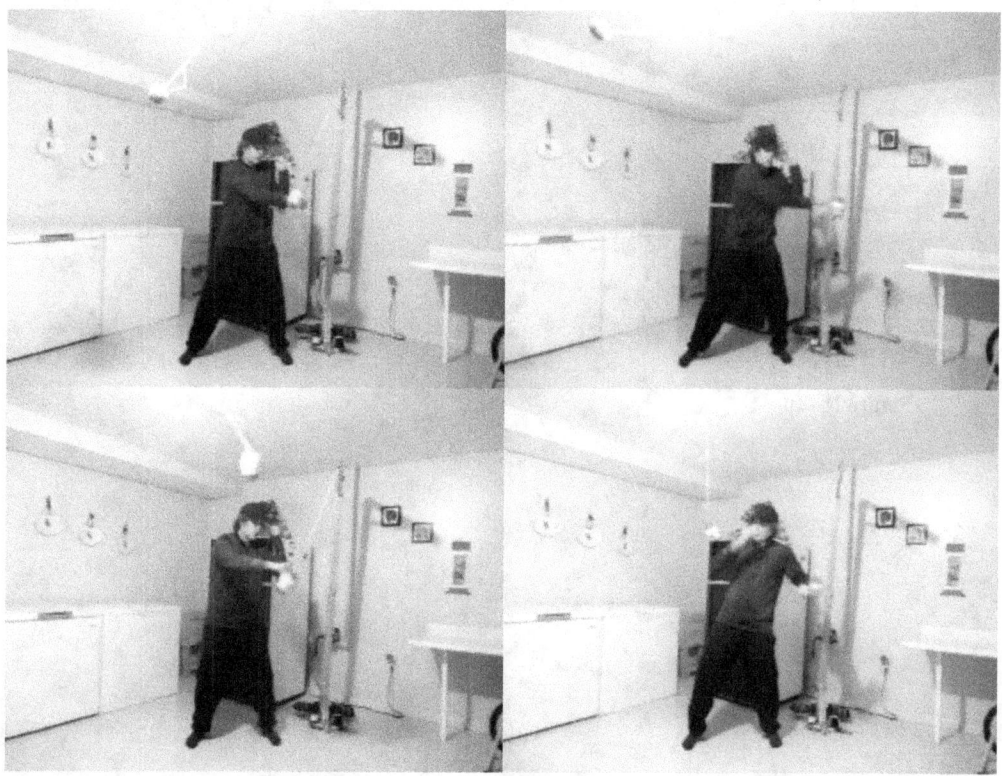

Figure 16- 9

You can combine punches, elbow strikes, knee strikes, blocking, and kicking for a greater and more fun workout, Figure 16-10. The goal is to let instinct take over so that you think less about movement and just react.

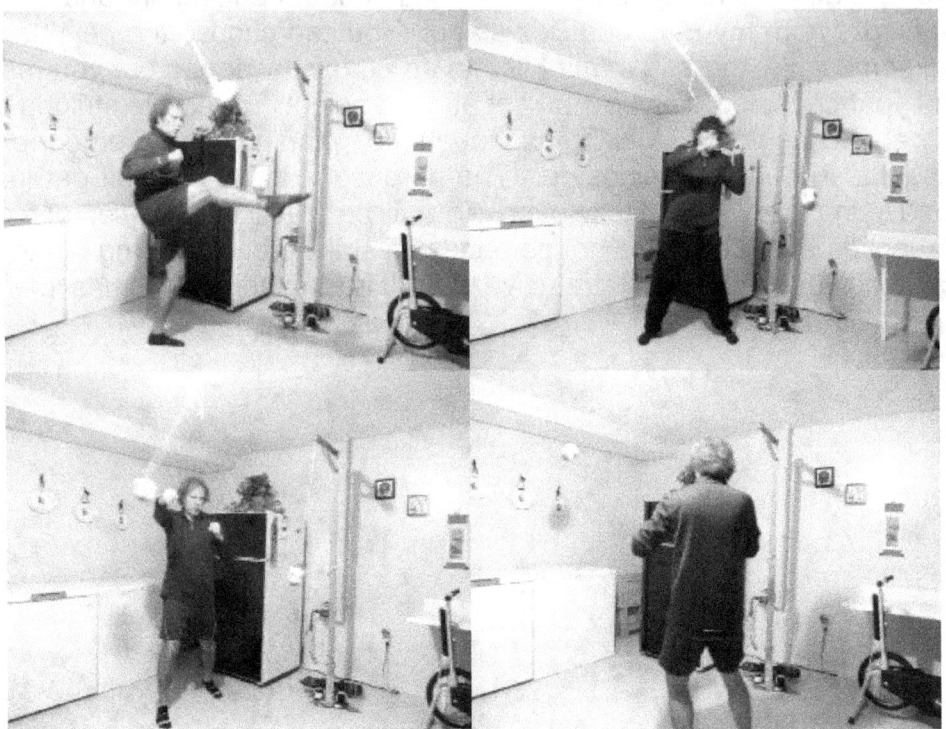

Figure 16- 10

Exercise with Bike and Box

Bike and Box is a mini punching bag apparatus that I invented that is designed to be used while biking on a low handlebar exercise bike, Figure 16-11. It is an excellent workout and motivator for anyone bored by exercise bikes.

Figure 16- 11

The bag hangs on a long cord allowing for varied directions for the bag to go. Basically it gives you something interesting to do while pedaling. It allows you to have a focus which is striking the bag and doing combinations of strikes while biking. You get extra calorie burn.

It also just happens to be a great exercise for Parkinson's. You must use arm movement, similar to boxing, only you also get to chase the bag.

You can do different punches, back-fists, speedbag strikes, elbow strikes. Holding arms up doing the strikes is a great workout out for the shoulders and back and allows you to work on fluidity and initiation of motion.

I also have punching exercises specifically for Parkinson's, tailored as warmups, and for times when Parkinson's meds are not at full strength, such as the karate punch and one hand on the handlebar strikes.

There are a variety of strikes that can be done. In addition there are even more combinations which keep the exercise interesting, almost addictive, while also encouraging flexibility and smoothness in arm motion.

The following are some of the strikes that are possible...

Karate punch

The karate punch, Figure 16-12, is punching in the style that karate punches are taught and which are frequently demonstrated in Katas, choreographed movements.

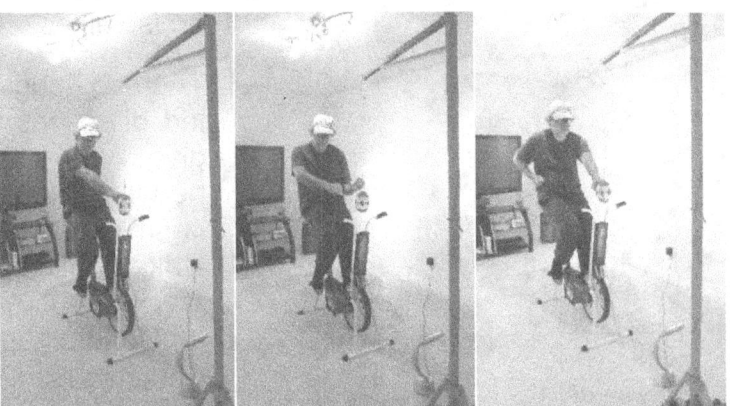

Figure 16- 12

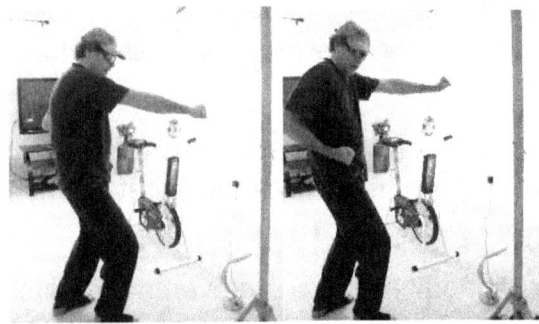

Figure 16- 13

Begin with the fist closed, palm side pointing upwards, and fist beside your waist. As you punch, twist your fist so that your knuckles are on top just before striking the punching bag, Figure 16-13. After punching pull back your fist, returning to the position beside your waist while simultaneously throwing the same kind of punching with the other arm.

If your meds are low this twisting motion can take extra effort until you get into a groove. To demonstrate the reason for doing this twisting strike, try the following. While sitting on your exercise bike, try holding fists in front of you and punching with fists perpendicular to the ground. Do a few times. Then try a few karate punches. You'll notice that your strike is harder on the bag with the karate punch. To do the twisting your brain sends an extra message to your arm, which opens up the neural pathway allowing your arm's initiation of movement to be smoother and stronger for the punch. You'll notice the difference when meds are not at full strength.

Wide arc punches

To do the wide arc punches you punch the bag forward and on its return, back-fist or speedbag-fist the bag to the left of center. Then as the bag travels on an arc around to the right side, creating a loop, punch the bag when it is far to your right. You should turn your upper body slightly to follow the bag as you strike, Figure 16-14. Repeat as the bag travels on an arc to the left side.

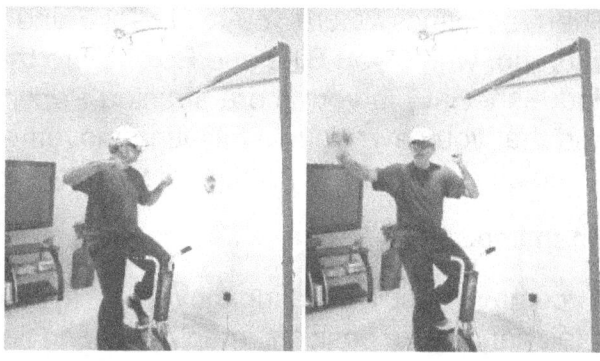

Figure 16- 14

When you get a rhythm going, it can be quite fun. You will eventually learn to transition into other combinations. Variety keeps this exercise interesting. Feeling that you are overcoming Parkinson's is another bonus.

Wide arc back elbow strike

The wide arc back elbow strike is done by sending the bag on an arc towards, for example, the right. To strike the right arm prepares by positioning the right arm, bent and in front of the body with the right fist in front of the left shoulder. As the bag loops toward the right ribs or shoulder, the right arm is moved in a way that makes the right elbow hit the bag, Figure 16-15

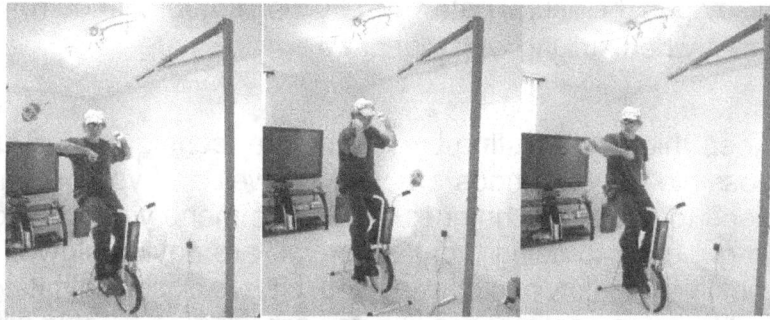

Figure 16- 15

Other strikes that can be done include:

- A boxing style strike punch
- A back-fist strike
- A Speedbag side-of-palm strike
- A forward elbow strike
- A karate chop

All these strikes can be combined into various combinations.

Techniques to help initiate movement

There are times where you might Bike and Box when meds are a little low or, like I often do, where you Bike and Box before taking your first meds of the day. You will find, especially in your more affected Parkinson's arm, that your punches are softer and that you have trouble initiating movement for the strike. There are a number of techniques you can rely on to help initiate movement.

Snap fingers

If you have trouble initiating movement in your left arm, then as the bag swings toward a left hand strike position, snap right hand fingers once or twice. This widens the neural pathway for a second or two. Your initiation of the strike will be smoother and with more force. Use for any left arm strike: a punch, back-fist, elbow, etc.

Open and close hand

This works the same as snapping the fingers only instead of snapping the fingers of your good hand, just open and close it just before striking with your weaker arm.

Alternate with karate chop

While doing repetitive speedbag style strikes using one hand, alternate between a closed fist speedbag strike and a karate chop. Opening and close your fist should also open the neural pathway, allowing smoother control and better initiation of movement when striking with that hand.

Keep things interesting by varying the techniques you use and by combining when possible. If your meds are working well and you are moving smoothly you will not need to use the techniques but I recommend you do practice them some of that time to get used to doing them when you are having slower moments. If Parkinson's is making you slow and causing you to have trouble initiating movement, the techniques will bring you satisfaction like you are winning a battle with Parkinson's.

Other techniques to help initiate punches (and kick):

Other techniques you can try in conjunction with punching are: smiling, chewing gum, head-mounted laser, music with a strong beat, wooden Neural Pathway Enhancers, Figure 16-16.

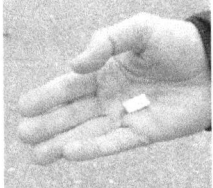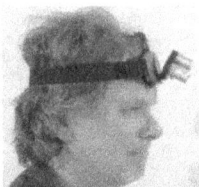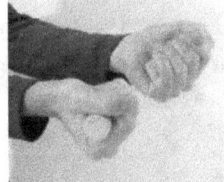

Figure 16- 16

Other good exercises for Parkinson's

- Cross country skiing
- Sit-ups
- Hitting a punching bag

Chapter 17

Medication Strategies

Levodopa is considered the most effective medication for Parkinson's. If you start on it you will feel great and almost normal all the time. But within five years a side-effect called Dyskinesia will likely have begun to be part of your life. It will continue to grow. Another called Dystonia will likely begin too. By year nine both side-effects will become a big part of your daily concern.

I am not a doctor so I speak from my experience. I believe you should delay taking Levodopa or Stalevo as long as possible, instead opting to try Dopamine Agonists or anything else first. If you do take Levodopa or Stalevo, try to minimize number of doses per day and to reduce dose size, even if the result is not perfect. For example, if you work 9 to 5, possibly you can get in the habit of not taking evening dose of Levodopa/Stalevo, or reducing the size of that last daily dose, if you are staying home during the evening. If your symptoms are manageable at home in the evening, I do recommend that you try this even if you have no Dyskinesia or Dystonia yet. It is the long game we are playing. Reduction will likely slow side-effects down. Believe me you will not like those side-effects once they become part of your daily life. If you get prescribed Levodopa or Stalevo, even reducing as I suggest you will still develop the side-effects over time but it might take the side-effects longer to grow.

Chapter 18

How I Learned to Experiment

Why I do this

I am not a doctor and I am not a scientist. I am someone who was diagnosed with Parkinson's in my early 40's. I followed my specialist's instructions and took my medication at the required times. Then in my fifth year a side-effect called Dyskinesia started happening.

When I finally saw my specialist again the meds which used to help for five hours were at that time wearing off more quickly, in addition I started getting Dyskinesia. My prescription of Levodopa was increased by 50%, from 100MG dose to 150MG dose to try to help my Parkinson's symptoms. It did help, however Dyskinesia really grew in my left leg and in my waist under that new prescription. Suddenly I had Dystonia growingg too.

I was prescribed Amantadine to help with the side-effects. On three doses a day of Amantadine I suddenly started having really vivid, really crazy nightmares nightly. I couldn't remember the last time I had a nightmare and suddenly I had multiple nightmares every night. I looked up Amantadine and sure enough it was linked to nightmares. So I decided to avoid the third dose which I had been taking around 5pm, hoping that avoiding the dose closest to my sleep time (11pm) would make the nightmares stop. I should say that taking Amantadine did help my Parkinson's symptoms and Stalevo side-effects. I still had side-effects (Dyskinesia and Dystonia) but they were decreased during the daytime. But I was taking three doses of Stalevo daily, the third dose at 5pm. Not taking the 5pm dose of Amantadine meant that Dyskinesia and Dystonia would be at full strength during my late evenings, coming on strong from around 9pm until 11pm disrupting my relaxation in front of the TV as Stalevo wore off. This meant my left leg would constantly twitch and jump, my torso would try to twist and lock up, my arms would stiffen out uncomfortably, my shoulders would tighten, and my head would be pushed/pulled forward at a time of the day when I really wanted to relax. There was no other side-effect medicine that I could take. It would be nearly a year before my next Parkinson's specialist appointment to see if I could try something other than Stalevo.

One reason patients get Deep Brain Stimulation surgery is due to the side-effects of Levadopa/Stalevo becoming unbearably strong. The benefit of DBS surgery is that it is hoped the patient can reduce and even stop taking those meds, effectively stopping the side-effects. I realized I had to reduce my daily intake of Stalevo, in particular was the need to reduce the 5pm dose. So this was the beginning of my experimenting.

Already an avid reader of articles on Parkinson's by this time, I started focusing more on news and forums about medication side-effects. I wanted to know experiences of other PD people, what seemed to help, what seemed not to help.

I also realized that Levodopa was causing my side-effects and that 50% increase in daily dose had sent it skyrocketing. So I decided to not take it for a day and sometimes for days. I would experience my Parkinson's symptoms which would come on stronger each additional day-in-a-row that I was off meds but surprisingly Day One was not bad. But what was truly great was that Dyskinesia went away while off Levodopa/Stalevo. Instead of spending an evening with my left leg involuntarily moving uncontrollably for hours as well as my torso writhing for the same period, I was instead able to sit peacefully watching evening tv. I moved around the house well without the evening dose so it became my practice and I would only take a 3rd dose if I was going out at night.

Three doses of 150 MG of Levodopa per day were leaving me with bad Dyskinesia from around 9pm until midnight. It was an unpleasant way to end each day. So I reduced to two doses a day if I was to be home during the evening. Now Dyskinesia came on strong from around 5:30 pm until 7pm each night. But it was less strong. After 7pm my body was at rest and it was relaxing watching TV, as it should be. At home in the evenings I found I moved well enough to not need that 3rd dose. I got into the habit of breaking pills to reduce the doses. Eventually I got my specialist to reduce my dosage to 125MG (also replacing Levodopa with Stalevo) and then back to the original 100MG. He did prescribe Amantadine and this helps with side-effects but only a little.

Over time Parkinson's advanced of course and there was only so many years I could hold out on two 100MG Stalevo per day.

I had also become focused on vitamins that were thought to be helpful to Parkinson's. The truth is that I don't know if these vitamins really benefit our Parkinson's brains but I have tried many and narrowed down the ones I prefer the most based on science and hope. If there is one we do need it is vitamin D3 because people with Parkinson's are low in Vitamin D.

I always incorporated exercise in my routine and it is known that vigorous exercise helps create additional temporary dopamine. I found cross-country skiing to be the best exercise for PD, at least until I invented Bike and Box (see *Chapter 16 - Exercise for Parkinson's*).

Eventually I slowed more and Dystonia grew despite the way I had slowed side-effects down. I did not like walking slow and luckily had bike paths through the forest to walk. I tried Nordic walking poles. They helped but when I slowed during the walk, the poles did not help me to go faster.

This led me to experimenting. I remembered how carrying something small in my left hand had helped me swing my left arm more normally when that arm's swing had been then only walking issue during the early days of PD. From there I went on to experiment on the techniques that I list in this book. Because I do take days off from Stalevo I do slow down enough times to frequently get to test my techniques on slow movement and trouble initiating movement.

Morning vs Evening

I noticed that in the first two hours of the day that I move well before taking my Parkinson's medication. I can go for a long walk using techniques or exercise with Bike and Box (Chapter 8). I take advantage of this as I believe it is good practice to keep your body used to moving without absolute reliance on meds. I've also trained my system not to expect meds in the evening. This gives me a break from Dyskinesia and Dystonia as well as helps slow the progression of the two side-effects. After the wearing off of my second dose of meds finishes, I am able to move around the house pretty well. I might go for a walk in the late evening some eight hours after my last dose of Stalevo. I actually realized that I walk best between midnight and 6am, possibly this is related to the same reason I have insomnia at night and sleepiness during the day, meaning Parkinson's makes me more 'on' at night.

Confidence

I hope these techniques make your days a little better and bring you more confidence and improved mobility.

Chapter 19

Conclusion and Final Thoughts

Every technique used, whether with a ball, chewing gum, using palm pressure, etc. increases the messages going through the neural message pathway making the muscles of our body respond and move better. It is my theory that the more often we practice these techniques that give a boost of 30 or 40% better use of our muscles, the better it also is for our digestive muscles too. Exercise, frequent use of my techniques, and a good diet should impact this part of our lives.

Even when seated watching television, a coin or something small can be held in your palm and in theory (my theory) the messages to our digestive system should be getting an extra boost during that time.

We Parkinson's people need every advantage to try to keep active in public once middle stage of PD has been reached.